The Economics of Mobile Telecommunications

The mobile telecommunications industry is one of the most rapidly growing sectors around the world. This book offers a comprehensive economic analysis of the main determinants of growth in the industry. Harald Gruber demonstrates the importance of competitive entry and the setting of technological standards, both of which play a central role in the fast diffusion of technology. Detailed country studies provide empirical evidence for the development of the main themes: the diffusion of mobile telecommunications services, the pricing policies in network industries, the role of entry barriers such as radio spectrum and spectrum allocation procedures. This research-based survey will appeal to a wide range of applied industrial economists within universities, government and the industry itself.

HARALD GRUBER is Deputy Economic Advisor at the European Investment Bank, Luxembourg, where he is responsible for project appraisal and sector studies in the information and telecommunications sectors. He has published extensively in refereed economics and industrial organisation journals and is author of *Learning and Strategic Product Innovation: Theory and Evidence for the Semiconductor Industry* (1994).

The Economics of Mobile
Telecommunications

HARALD GRUBER

CAMBRIDGE
UNIVERSITY PRESS

CAMBRIDGE UNIVERSITY PRESS
Cambridge, New York, Melbourne, Madrid, Cape Town, Singapore, São Paulo

CAMBRIDGE UNIVERSITY PRESS
The Edinburgh Building, Cambridge, CB2 2RU, UK

Published in the United States of America by Cambridge University Press, New York
www.cambridge.org
Information on this title: www.cambridge.org/9780521843270

First published 2005

Printed in the United Kingdom at the University Press, Cambridge

A catalogue record for this book is available from the British Library

Library of Congress Cataloguing in Publication data

Gruber, Harald.
　The economics of mobile telecommunications / Harald Gruber.
　　p.　cm.
　Includes bibliographical references and index.
　ISBN 0-521-84327-8 (hb : alk. paper)
　1. Cellular telephone services industry.　2. Mobile communication systems – Economic
aspects.　3. Wireless communication systems – Economic aspects.　I. Title.
　HE9713.G78 2005　2004054225
　384.5′33–dc22

ISBN-13 978-0-521-84327-0 hardback
ISBN-10 0-521-84327-8 hardback

For Licia

Contents

Figures

Tables

Preface

This book distils years of work on the mobile telecommunications industry. I became interested in this industry for professional reasons during the mid-1990s, a period when the industry was making the jump from a premium service industry for mostly professional users to a truly mass market. In my capacity as an applied industrial organisation economist, I had the unique opportunity of evaluating the business plans and strategies of a large number of mobile telecommunications firms inside and outside Europe. This provided me with valuable insights into the functioning of this fascinating industry, as well as into its technological and operational concerns.

This book makes extensive use of previously published material. It thus also benefits from joint work done with Marion Hoenicke, Tommaso Valletti and, in particular, Frank Verboven. The credit to them is given in the appropriate sections throughout the book and the relevant papers are quoted in the bibliography. Researching and writing articles with all of them was an intellectually very rewarding experience, and I owe them my thanks. I also received many useful comments and hints from colleagues within the EIB and from the academic world. I would like to thank Tommaso Valletti and two anonymous referees for having read the manuscript and for their detailed comments. Ultimately, all responsibility for the views expressed remains with the author, and they do not necessarily reflect those of the European Investment Bank.

Abbreviations and acronyms

Telecommunications terms

AM	Amplitude modulation
AMPS	Advanced mobile phone service
ARPU	Average revenue per user
C-450	German analogue mobile standard
CAMEL	Customised application mobility enhanced logic
CCIR	International Radio Consultative Committee
CDMA	Code division multiple access
CDMA 2000	A 3G system based on CDMA
CEPT	European Conference of Postal and Telecommunications Administrations
CLEC	Competitive local exchange carrier
CPP	Calling party pays
CTIA	Cellular Telecommunications Industry Association (US)
D-AMPS	Digital AMPS = US-TDMA
DCS 1800	Digital communications system = GSM 1800
DECT	Digital enhanced cordless telephony
EDGE	Enhanced data GSM environment; also 2.5G
ERC	European Radio Communications Committee
ERO	European Radio Communications Office
ETSI	European Telecommunications Standardisation Institute
FCC	Federal Communications Commission (US)
FDMA	Frequency division multiple access
FM	Frequency modulation
FTM	Fixed to mobile
GPRS	General packet radio service; also 2.5G
GSM	Global system for mobile communications (formerly *Groupe système mobile*)
GSM 900	GSM in the 900 MHz band
GSM 1800	GSM in the 1800 MHz band = DCS 1800
GSM 1900	GSM in the 1900 MHz band = PCS 1900

HLR	Home location register
HSCSD	High-speed circuit switched data
iDEN	Integrated digital enhanced network
IMT-2000	International mobile telecommunications system: the ITU definition for 3G
IMTS	Improved mobile telephone service
IOT	Inter operator tariff
IS 95	Interim standard (US) describing the CDMA air interface
IS 136	Interim standard (US) describing the D-AMPS air interface
ITU	International Telecommunications Union
JDC	Japanese digital cellular = PDC (Japanese digital mobile standard)
JTACS	Japanese TACS
MSC	Mobile switching centre
MTF	Mobile to fixed
MTM	Mobile to mobile
MVNO	Mobile virtual network operator
NMT	Nordic mobile telephony system (in 450 and 900 MHz bands) (Scandinavian analogue standard)
NTT	Nippon Telephone and Telegraph Cellular System (Japanese analogue mobile standard)
Oftel	Office of Telecommunications (UK)
ONP	Open network provision
PCN	Personal communications network (UK) operating at 1800 MHz = GSM 1800
PCS	Personal communications services (US, Japan) operating at 1900 MHz
PDC	Personal digital cellular (Japanese digital mobile standard)
PHS	Personal handy phone (Japanese cordless system)
PLMN	Public land mobile network
PMR	Private mobile radio
PSTN	Public switched telephone network
RBOC	Regional Bell operating companies (US)
RC 2000	*Radiocommunication 2000* (French analogue mobile standard)
RPP	Receiving party pays
RSA	Rural Statistical Areas (US)
RTMS	Radio telephone mobile system (Italian analogue mobile standard)
SIM	Subscriber identification module
SMS	Short message service
SNR	Signal-to-noise ratio

TACS	Total access communications system (an analogue mobile standard)
TDD	Time division duplex
TDMA	Time division multiple access (also D-AMPS)
TD-SCDMA	A 3G system based on CDMA
TETRA	Trans-European trunked radio communications
TIA	Telecommunication Industry Association (US)
UMTS	Universal mobile telecommunications system
UTRA	UMTS terrestrial radio air interface
VLR	Visitors' location register
VPN	Virtual private network
W-CDMA	Wideband CDMA (the basis for UMTS)
WRC	World Radiocommunication Conference
1G	First-generation (analogue) cellular technology
2G	Second-generation cellular technology
2.5G	Enhanced 2G (GPRS, EDGE)
3G	Third-generation cellular technology

General terms

ANSI	American National Standards Institute
BTA	Basic trading areas (US)
CAGR	Compound annual average growth rate
CEE	Central and Eastern Europe
ECPR	Efficient component pricing rule
EMU	European Monetary Union
GDP	Gross domestic product
ITC	International Trade Commission
JV	Joint venture
LRIC	Long-run incremental cost
M&A	Mergers and acquisitions
MoU	Memorandum of Understanding
MSA	Metropolitan Statistical Areas (US)
MTA	Major trading areas (US)
PPP	Purchasing power parity
R&D	Research and development
ROCE	Return on capital employed

1 Introduction

1.1 A new and fast-growing industry

A series of features makes the mobile telecommunications industry an interesting field of investigation for economists: the industry is experiencing very fast market growth combined with rapid technological change; regulatory design in setting market structure is playing a very important role; and oligopolistic competition is unfolding under various forms. The number of subscribers to mobile networks is growing at a rapid rate on a worldwide basis, as shown in figure 1.1. During the 1990s the number of mobile subscribers worldwide increased by an annual rate of 50 per cent. An important year was 2002, when the number of world mobile subscribers for the first time exceeded the number of fixed lines. The number of mobile subscribers was close to 1.2 billion at the end of 2002, while the number of fixed lines was slightly below 1.1 billion. The year 2002 therefore established at worldwide level what had already been observed for an increasing number of countries during the previous few years: mobile telecommunications is the most widespread access tool for telecommunications services. The mobile telecommunications industry has acquired as many users in some twenty years worldwide which took the fixed line telecommunications industry more than 120 years to achieve.

The timely and efficient supply of mobile telecommunication services has had a substantial impact on the economy, which also explains the extensive public interest in this industry. The actions of the industry regulator are of crucial importance for this. For instance, a study on the US market shows that the regulatory delay in licensing mobile telecommunications gave the US consumers welfare losses in the range of $24–50 billion a year.[1]

As will be shown in this book, two factors have determined the extraordinary rapid development of this industry: *technological progress* and

---

[1] This figure is quoted from Hausman (1997). However also other studies such as Rohlfs, Jackson and Kelley (1991) find such orders of magnitudes.

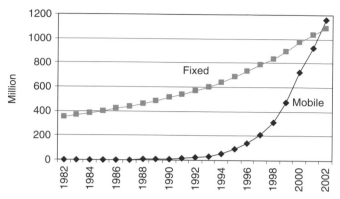

Figure 1.1 *The evolution of the worldwide number of mobile and fixed telecommunications lines, 1982–2002*
Source: ITU data.

regulation. The mobile telecommunications industry as it is known today – i.e. using radio waves instead of wires to connect users – is a relatively young industry. However, its basic technological concepts actually date back to the second half of the nineteenth century, when the German scientist Heinrich Rudolf Hertz demonstrated (in 1888) that an electric spark of sufficient intensity at the emitting end could be captured by an appropriately designed receiver and induce action at a distance. The first mobile telecommunications systems were based on the same principles as radio or television broadcasting, by which all conversations could be heard by everybody. These systems had very limited capacity and used the electromagnetic radio spectrum, whose usable portion is only very limited, in a very inefficient way. Significant progress in using the spectrum more efficiently and ensuring privacy in conversations were made with the development of the 'cellular' concept after the Second World War. However it took until the 1970s for the progress in semiconductor technology to allow the construction of cellular mobile networks for commercial use. Analogue technology cellular systems were introduced first at the beginning of the 1980s. The breakthrough for a mass market for mobile telephony occurred only in the 1990s with the advent of digital technology. The scarcity of radio frequencies, necessary for transmission between the user's handset and base stations, has since then constituted the bottleneck for the development of the industry. As we have seen, the early analogue technology used the allocated radio frequency spectrum in a relatively inefficient manner so only a relatively small number of subscribers could be connected, who used the system mainly for business purposes. The introduction of digital technology led to a breakthrough in performance, capacity and quality of mobile telecommunications. Digital technology, such as the European standard, GSM, made better use of the radio spectrum than analogue technology did and could therefore accommodate more subscribers. Lower unit costs could be achieved by spreading fixed costs over more subscribers.

Regulatory reform is the other driving force behind the spreading of mobile telecommunications. Because of the radio spectrum constraint, the industry is structurally considered as an oligopoly and the development of the industry crucially depends on pre-entry regulation. In emerging industries, characterised by significant technological progress, there is usually little consensus on the optimum policies concerning the development of the sector. Among other issues, the debate focuses on how and when entry should be promoted and whether technology standards should be imposed centrally or selected by the market forces in a decentralised way. Because of the lack of consensus, governments have taken different policy options, and often change directions as experience accumulates.

The effects of entry in the cellular mobile industry are particularly interesting to analyse. Radio spectrum is the scarce resource to be assigned and constitutes the entry barrier for the firms. However, technological progress permits greater efficiency in spectrum usage and thus potential for accommodating more firms. Governments throughout the world have also taken quite different options regarding the timing and the number of entry licences. This provides interesting data for assessing the effects of licensing on the evolution of the industry.

Such pre-entry regulation in mobile telecommunications has various dimensions. First, the policy maker needs to decide whether to set a single national (or international) standard, or whether to allow multiple technological systems to compete. Second, the policy maker has to decide how many firms a licence will be granted. This also involves an important decision with respect to the timing of first and additional licences. Third, the government needs to decide how to grant licences. In the early days of mobile telecommunications, licences were often granted on a first-come-first-served basis. With the introduction of the cellular technology, the first licences were frequently granted by default to the incumbent fixed operators. Additional licences were initially granted through an administrative tender procedure (lotteries, or 'beauty contests') and then more and more through auctions. This evolution has greatly changed the nature of the firms in the market and their competitive behaviour.

Economic theory can give guidance on these issues, but the propositions of traditional textbook economics are complicated by the fact that mobile telecommunications is a network industry. For instance, in markets without network effects, it seems to be unambiguously desirable to allow multiple competing technological systems. In contrast, in markets with network externalities there are both advantages and disadvantages to having multiple systems rather than a single standard. The presence of (strong) network externalities typically leads to 'tipping' markets, where the winning technology takes the whole market. Should the government intervene in this race by imposing a single standard? Or should the markets decide themselves on which standard will eventually 'win'? The theoretical literature does not provide an unambiguous answer to these questions.[2]

There is also the question to which extent network externalities are in fact present in cellular telecommunications markets. The main sources of network externalities arise from the fact that mobile users can use their handset only within the areas that support their technological system. Thus, depending on the mobility of consumers, network externalities are local, national, or even international in scope. In addition to reducing consumer switching costs and creating 'roaming' possibilities, the presence of a single technological system also has the traditional advantage of exploiting economies of scale in the manufacture of equipment. Various incompatible technological systems have been developed in the cellular mobile telecommunications industry (most of them with the support of leading countries). Each system is subject to network externalities in that consumers value a system more the more users adopt it. The relevant policy question is whether governments should impose a single standard, or whether the markets should select a winning standard in a decentralised way. Advantages of mandatory standards are that potential network externalities can be realised faster, and that users' technological uncertainty is reduced. Advantages from a decentralised approach are that there may be less a risk of being 'locked in' with inferior technologies and that incentives for innovation to better systems are preserved. Yet a counterargument is that also the decentralised, market-based, approach may lead to lock-in with inefficient technologies. Despite the extensive theoretical literature, there exists little empirical work that compares the effect of imposing standards on the diffusion of a new technology with the effect of allowing multiple systems to compete. Again, the cellular mobile telecommunications industry offers an interesting opportunity to make such a comparison, since countries have followed quite different and changing policies regarding standards. While chapter 2 gives a general overview of the main issues affecting the mobile telecommunications service industry, chapter 3 is an extensive description of the evolution of the mobile telecommunications industry looking at representative countries. The aim is

---

[2] See, for instance, Katz and Shapiro (1994) and Shapiro and Varian (1999).

to highlight the importance of country-specific effects, especially at the beginning of the industry. These country-specific effects tend to peter out as the industry progresses. Chapter 4 provides answers to questions of the role of different regulatory policies on the diffusion of cellular mobile telecommunications, relying on quantitative methods and using a world-wide data set.

1.2 Business strategies for firms

One of the main features of a mobile telecommunications network is to provide *coverage*. The fact that a user can utilise a mobile phone over a very large portion of the territory distinguishes it from the fixed network. This coverage can be provided by only a limited number of firms. The radio spectrum bottleneck acts as barrier to entry and makes the industry intrinsically oligopolistic. The question arises which type of strategies firms are able to pursue in such an environment concerning pricing and product positioning. For instance, there may be scope for vertical product differentiation by providing different levels of coverage. However, differentiation in coverage seems to be possible to only a limited extent, mostly during the early years of the life cycle of the industry, when firms have to spread network build-out over time for cost reasons, but in the longer term firms typically have regulatory obligations to provide full coverage. This means that there is little scope for relaxing price competition through product differentiation in terms of coverage. But when differentiation is possible, studies shown that price competition is relaxed. Empirical studies also show that price competition is of the Cournot type, i.e. with price above marginal cost and decreasing with the number of firms in the market.

Pricing of mobile telecommunications services is multidimensional and hence complex, both at the wholesale and the retail level. Retail pricing decisions concern mainly services such as subscription, on-net and off-net calls. Wholesale pricing also include interconnection pricing among networks. Theory provides limited guidance, as the economic literature still has to explore many aspects of pricing in network industries. The market power of individual firms may be exerted to a different degree at each level. It may thus be important from a social welfare point of view to check abuse of market power through 'ex ante' regulation – i.e. through measures that limit damaging behaviour before it occurs. There is a consensus among the policy makers that such 'ex ante' regulation, if necessary at all, should be as light as possible. This implies that such regulation should be much lighter in mobile telecommunications than in fixed telecommunications, where 'natural monopoly' positions seem to be much more entrenched.

Regulators took some time to appreciate that cost allocation mechanisms could be profoundly different between fixed and mobile networks. While fixed network infrastructure used to be based on plant and equipment that from an accounting point of view had been depreciated, mobile network infrastructure was typically new and thus carried high depreciation charges in cost accounting. This, for instance, led to regulated interconnection prices that were favourable to mobile telecommunications firms. Cost allocation mechanisms are important when it comes to establishing other aspects of interfirm compensations and how these are transferred to the users. There are two principles: calling party pays (CPP) and receiving party pays (RPP). Although from a theoretical point of view RPP seems to have better characteristics for ensuring allocative efficiency, CPP has been the overwhelming success in terms of worldwide diffusion. Only a few countries, in particular the USA, actually have RPP in place, and for legacy reasons rather than for choice. CPP allows firms to exercise market power in call termination. The favourable interconnection arrangements with CPP provided the mobile telecommunications industry with the financial resources for subsidising the acquisition of customers, and this may account for a substantial part of the rapid growth in the mobile telecommunications subscriber base. Regulatory attempts are underway to fence in the market power mobile telecommunications firms have on traffic termination. Similar considerations apply for international 'roaming', where there are actually elements of RPP but where firms are nevertheless able to exploit the lack of information on the customer side. In any case, the evolution of overall mobile telecommunications service pricing shows a general trend towards more competitive pricing, but there are still some large areas where this does not apply. These issues are addressed in detail in chapter 5, which sets a framework for the business strategies concerning product positioning and pricing. Particular attention is devoted to market segments where market power can be exercised more easily.

1.3 Radio spectrum availability as a key determinant for market structure

Radio spectrum, the key input for the supply of mobile telecommunications services, is a public good, but its use is exclusive when employed for mobile telecommunications services. Its allocation thus needs to be regulated. Other services such as broadcasting compete for the allocation of spectrum and hence only a limited portion of the spectrum is available for mobile telecommunications services.[3] This combined with the high sunk

---

[3] The technical properties of the radio spectrum and the technical description of mobile telecommunications are discussed in more detail in the appendix.

costs for the set up of mobile telecommunications networks leads to the consequence that the market can support only few firms. The frequency assignment mechanism to firms is important for two reasons: first, radio frequencies are a scarce resource for exclusive use; second, radio frequencies provide a potential for oligopoly rents. For these reasons, the spectrum assignment method is very important and can be divided into two major categories: administrative methods, such as 'beauty contests', and market-based methods, such as auctions. There are refinements for each category, but the main difference boils down to the role of information retained by the government. With administrative methods the government plays an active part in the assignment and advocates a central role in the development of the industry. With a market-based method the government prefers to compare itself with a referee, setting the framework and letting firms decide on the implementation of measures for the development of the market. For instance, the auction mechanism is based on the belief that the market has sufficient capability for self-selection to award spectrum to the firms that make the most efficient use of it, and that this is in the public interest. Chapter 6 surveys the different assignment mechanisms, presenting their advantages and disadvantages. The experiences of selected countries are documented in some detail to illustrate these points. The most important episode in this respect is the assignment of so-called 'third generation' (3G) licences in Europe. This has shown that auctions in general deliver much higher receipts to the governments than do administrative methods. Moreover, the design of auctions, in particular with the aim of avoiding collusion, is of utmost importance in generating large receipts. However, serious doubts have arisen on whether bidding agents are really better able than governments in assessing market prospects.

Although entry into the mobile telecommunications market is regulated, there is the question whether the industry is a 'natural oligopoly'. If spectrum were not a scarce resource, other factors, such as sunk costs or scope for vertical product differentiation, could set in as determinants of market structure. The historically observed evolution of market structure in the mobile telecommunications industry is from higher to lower levels of concentration. In most countries, the industry has evolved from a monopoly to an oligopoly with three or more firms. Waves of generations of technology have typically been a trigger for additional entry, as newer generations of technology with more efficient use of radio spectrum permitted the entry of more firms. This entry has been sequential, and the profitability of the industry has declined, with new entrants being less profitable than long-established firms. The question now arises of whether entry has led the industry to the zero profit level. This could be indicated by the observed exit or attempts to merge of late entrants, as being noted in some countries. This is particularly relevant in the forthcoming market for 3G mobile services in Europe, where a new design of market structure has taken place. Governments decided for simultaneous entry of a larger number of firms than for 2G (second-generation) mobile services, but with apparently little assessment of whether the new market would support such a large number of firms. Moreover, there has been a tendency to privilege auctions as the assignment method. It has turned out that with auctions there is a tendency to increase the number of firms in the industry, and with individual firms paying more than with other assignment methods. Chapter 7 develops a benchmark model that illustrates the interplay between sunk costs, such as licence fees, and market structure. It suggests that 'overbidding' of licence fees may occur, at the expense of forsaking the market structure envisaged by the policy maker or of collusion at the post-entry stage in the market. The model's predictions are compared with the outcomes from 3G licensing in Europe and subsequent events. Evidence of exit of firms and calls for relaxation of licence conditions suggest that overbidding had taken place: even in cases with zero licence fees the government has apparently allowed for too much entry. This may suggest that the industry has arrived at a point where spectrum is no longer a constraint. Entry may not even need regulation any more. If this were to be the case, it would mark the emancipation of the industry from the spectrum bottleneck.

2 Stylised features of the mobile telecommunications industry

2.1 Introduction

Mobile telecommunications use radio waves,[1] instead of wires, to connect users. Though the origins of wireless communications may be traced back to the second half of the nineteenth century, the earliest applications for mobile communications date back to the 1920s. After the Second World War, when the civilian use of wireless telecommunications resumed, several industrialised countries independently developed mobile telecommunications systems. These, however, suffered of a series of technical limitations that hampered their widespread use. Only during the 1980s did these problems begin to be surmounted, with the diffusion of cellular mobile telecommunications technology as it is known today. To fully appreciate the technological challenges mobile telecommunications had to surmount to become a widely spread technology, it is useful to briefly sketch the history of the technology in the context of the working principle of wireless communications. This chapter outlines the main driving forces of the mobile telecommunications industry and how they shape the evolution of the sector and gives some hints on the prospects for the future of the sector. The key issues will be dealt with in more detail in subsequent chapters. This chapter is organised as follows. Section 2.2 presents a brief history of the technological developments in the mobile telecommunications industry. Section 2.3 provides some notions of the different technologies available. Section 2.4 illustrates some of the main user trends in this fast-growing industry, while section 2.5 looks at the revenue side. Section 2.6 takes a closer look at the cost side, which proves to be very important in driving penetration of mobile telecommunications: even

---

[1] Radio waves are a natural resource and only a small part of the total electromagnetic spectrum is suitable for radio transmission. The measurement unit is Hertz (Hz) which indicates the cycle per second. For more technical details, the reader is referred to the appendix.

though most of the cost elements are declining in this industry, some are increasing and could be of crucial importance. Section 2.7 discusses the main issues concerning regulation, both the pre- and post-entry stage. Section 2.8 draws some brief condusions.

2.2 Some technology history

2.2.1 Mobile telecommunications before the cellular era

The first attempts at wireless communications

The origins[2] of wireless communications may be traced back to the German scientist Heinrich Rudolf Hertz, who demonstrated in 1888 that an electric spark of sufficient intensity at the emitting end could be captured by an appropriately designed receiver and induce 'action at a distance'. This transmission via the 'ether' challenged the classical notions of physics. Whereas Hertz's experiments spanned just a few metres, it was the Italian scientist Guglielmo Marconi who constructed a 'radio' that transmitted waves over increasing distance: In 1895, he transmitted signals over a distance of 2.5 km, in 1899 over the English Channel and in 1900 over more than 300 km.[3] Marconi's greatest challenge was to confute the conventional belief that radio waves propagated only linearly and therefore would be unable to follow the curved surface of the earth. In 1901, Marconi established the first wireless transmission over the Atlantic, spanning over 3500 km from Cornwall to Newfoundland. Maritime applications become the dominant market for wireless, even though only large and expensive ships could carry the wireless equipment and justify the cost.

At the beginning, only gross pulses of energy could be transmitted, and communications was limited to Morse code. Technological improvements, in particular the refinements in radio communication technology such as amplitude modulation (AM)[4] and the invention of the thermo-ionic valve, led to the possibility of transmission of speech and music. However, wireless equipment was a low-volume and high-cost market. Before the start of the First World War there were some 2000–3000 wireless in use in the entire world, most of them in Britain. At the outbreak of the war in Europe the development of wireless was intensified, again mostly for

---

[2] Historical accounts of the industry can be found in Calhoun (1988), Mehrotra (1994) and Garrard (1998), who also refer to primary sources.

[3] In 1896, Marconi offered his wireless system to the Italian government, but he never received a reply and eventually he decided to emigrate to England. There, he met Sir William Preece, the chief engineer of the telegraph office, who provided Marconi with the funds for an experimental site at Lavernock in Wales UK.

[4] Undertaken the first time by Reginald Fessenden in 1905, with AM information transmitted by varying the amplitude of radio waves.

maritime applications. The applications of wireless for ground-troop applications were still met by scepticism from the military planners, and also because of the bulkiness and weight of the equipment.

The first voice transmission

The surplus supply of valves after the First World War led professionals and amateurs to experiment with voice transmission. These first experiments were called 'voice telephony', even though they did not imply any aspects of switching and connectivity as expected today from telephony systems. The main drawback for commercial applications of these communications systems was the lack of privacy, since it was easy to eavesdrop on any conversation. But this drawback was turned into a benefit with the advent of broadcasting. Broadcasting enjoyed spectacular growth in the USA: in less than three years after the opening of the first broadcasting station in 1920 there were 500 stations with 2 million listeners. In 1924, there were 1100 stations and the unregulated use of radio frequencies led to chaos. In 1927, a first attempt was made to regulate spectrum usage during an international conference in Washington, it was agreed to allocate the frequency band from 550 kHz to 1.5 MHz to broadcasting. Frequencies below this were allocated to maritime communications. Europe was lagging behind in the evolution of broadcasting, which had been restricted much earlier through the set-up of public companies such as the British Broadcasting Corporation (BBC). There were some 200 broadcasting stations in Europe in 1929 and broadcasting transformed the wireless industry into a high-volume and low-cost industry, especially through the market for wireless radio receivers. Until the outbreak of the Second World War, the most important technological developments were made in this field.

The first attempts at true mobile date back to the early 1920s. In the USA in 1921, the Detroit Police Department made the first experiments with 'mobile' radio (Noble, 1962). At the beginning, the service was limited to a sort of paging, instructing the police car in question to stop and call back to the police station. These one-way systems were widely used in the USA. Similar experiments were carried out by the Metropolitan Police of London, though with less satisfactory results (Garrard, 1998). In 1932, the Brighton police force was equipped with radio equipment weighing just over 1 kilo, so that they could be carried by patrolling police officers. One-way messages could be sent to all officers within the range of 6 km.

A few years later proper two-way communication features were put in place, but for reasons of weight the equipment could be fitted only to vehicles. The British police, however, were reluctant to introduce these voice communications systems because of lack of privacy. The police preferred to fit its cars with telegraphy systems during the mid-1930s, telegraphy had the advantage of greater reach (up to 100 km) and was less likely to be eavesdropped, as relatively few persons were able to read Morse signals. The US police was less concerned with privacy, and preferred to adopt voice communication systems. Sweden also equipped its police force with two-way voice communication systems in the late 1930s. Most of the police radio systems at that time worked in the 1.5–3.0 MHz band, which at that time was found to give the best compromise between availability, interference and performance.

Private mobile radio

Radio communications played a vitally important role during Second World War operations, in particular in the air and at sea. By the end of the war, the army, especially the US army, was also extensively fitted with two-way equipment. At that time the US electrical equipment manufacturer Motorola coined the term 'Walkie-Talkie' for its two-way mobile radio. This used frequency modulation (FM[5]) instead of AM, thereby reducing weight and size of the equipment, while performance improved. Europe was much slower in adopting FM for mobile radio, and in countries such as Britain, mobile radio continued to use FM until well into the 1970s.

As the US army was exclusively using FM equipment, all radio communication manufacturers were geared to the production of FM systems. At the end of the Second World War, these manufacturers were looking for civilian applications, this gave the US equipment manufacturers a head start in the further development of mobile communications systems. These bi-directional FM systems became very popular and were sold mainly to public service organisations such as police, emergency services and taxis, as well as public utilities for water, gas and electricity. These systems are referred to as 'private mobile radio' (PMR), and constitute a *closed communications network* for a group of users who needed to stay in contact with a central controller, or dispatcher, and sometimes with each other, in which case connection is usually controlled by the dispatcher. PMRs were owned and operated by the organisations that used them and were not allowed to carry third-party traffic. PMRs were not interconnected with the public fixed telecommunications network.

The working principle of these early PMR systems is that an emitter is set up to cover as large an area as possible, in a very similar way to radio broadcasting. Each frequency channel is dedicated to a specific user. The drawback is that only relatively few users can talk at the same time and a

---

[5] With FM, information is transmitted by frequency modulation, instead of amplitude modulation (AM), this not only increases the quality of the sound, but also decreases the spectrum requirements and opens up the use of higher frequencies.

single frequency channel had to be assigned to each. Hence there is a high probability of a user being blocked in using the mobile telephone. To meet the growing demand for mobile telecommunications more frequencies had to be assigned to these services, but this was difficult since alternative uses of the same frequencies, such as radio and TV broadcasting, seemed to be socially more useful.

Two innovations were made to improve efficiency in the usage of the frequencies. One was the *splitting of channels* – i.e. the introduction of technologies to support voice transmission with a smaller bandwidth, so that with the same frequency band more users could be supported. The second innovation was 'trunking'–, i.e. making the full range of channels available to each individual user, instead of giving each user a dedicated channel.[6] This helped to reduce the probability of being blocked in making a call. Initially trunking was manual – i.e. each caller had to search through the available channels manually, determining by listening which channels were occupied and selecting an unused channel for the call. Later, trunking was performed automatically. Initially, dialling was through an operator; only during the mid-1960s, with the introduction of the 'improved mobile telephone service' (IMTS), was dialling automated in the USA. IMTS became the direct technical precursor of, and in some ways the prototype of, cellular radio.

Early pre-cellular mobile telecommunications systems had very limited capacity since they made use of the spectrum in a very inefficient way. The available portion of the radio frequencies in the overall spectrum is limited by both technology and regulation. Since there are many alternative uses for the radio spectrum (such as broadcasting or military applications), firms in the industry had difficulties in convincing governments to allocate a significant portion of the spectrum to mobile telecommunications.[7]

These early mobile radio systems were based on the same principles as radio or television broadcasting. They made use of high-power transmitters located in base stations on top of the highest point in the coverage area. The transmitters operated at very low frequency levels of around 150 MHz. At such low frequencies, signals travel very far, so that a base station has a large coverage, with a radius up to 80 km. This has the advantage that only few base stations are required to cover a geographic area. However, at the same time, the few available frequency channels to support telephone conversations are locked up over a large area and can

---

[6] The concept of trunking may be illustrated by the following example. If a channel in a system without trunking permits access to only two or three users per channel, with a likelihood of congestion not exceeding say 10 per cent, then the total capacity of a system with twenty channels is approximately fifty users. In a network with trunking, the number of users would increase to 420, with the same probability of blocking, because a subscriber could use any free channel. This advantage is called 'trunking efficiency'.

[7] Kargman (1978) and Levin (1971) provide a full description of these lobbying activities.

thus serve only a small number of users. In 1970, the Bell System in New York could support just twelve simultaneous mobile conversations, the thirteenth caller was blocked.

To meet the growing demand for mobile telecommunications more frequencies had to be assigned to these services, but this was difficult since as we have seen alternative uses of the same frequencies such as radio and TV broadcasting seemed to be socially more useful. PMR failed to have wide diffusion in Europe because of the relatively high cost and the limited use as a communications device beyond restricted user groups such as the police and other public services.

Technological progress in the equipment industry, such as the adoption of transistors in the mobile terminal during the early 1960s, helped to make the device portable, but ultimately did not trigger substantial further growth. PMR penetration reached the highest level in the USA, with 2.7 users per 100 inhabitants in 1977 (Garrard, 1998). In comparison, Sweden, as the most advanced European country in this field, reached a penetration rate of 1.6 users per 100 inhabitants by the 1960s.

By the 1960s the development of different wireless system created competition for spectrum: PMR[8] was competing not only with broadcasting and military use, which jointly accounted for more than two-thirds of frequencies below 1 GHz, but also with aviation, maritime, space and amateur applications. European regulatory authorities had quite different approaches in allocating the frequencies. The Scandinavian countries adopted a forward looking and commercial approach, and reserved more spectrum for mobile communications, an approach that was very useful in the launch of cellular mobile telecommunications. Other countries, such as the UK, had a more administrative approach that paid less attention to commercial issues and effective demand from the market. However, in all cases PMR never really became a widespread technology. Apart from the availability of frequencies, there are also other reasons for this. PMR is in principle based on operational control, and therefore applications never went beyond the closely defined purpose, PMRs were also often not interconnected with the fixed telecommunications network, mainly for regulatory reasons.

Diffusion of the first mobile telephones
USA The first true mobile telephone that was also interconnected with the fixed telecommunications system was introduced in the USA in 1946. The FCC granted a licence to AT&T to operate such a network in

---

[8] Most European countries allocated PMR frequencies in the VHF range (70–86 MHz, 104–108 MHz and 165–170 MHz), as well as in the ultra high frequency (UHF) range (425–462 MHz), although these were unsuitable for a wider use by commercial organisations.

St Louis. Within a year, the service was being offered in more than twenty-five USA cities. Mobile penetration reached the highest level in the USA, with 2.7 users per 100 inhabitants in 1977 (Garrard, 1998). This penetration was helped by a very liberal licensing policy adopted by the Federal Communication Commission (FCC) and by what market observers was considered as the generally technology friendly approach of US consumers.

Sweden The first mobile telephone system in Europe was launched in Sweden.[9] Since this can be considered as the pioneering country for mobile telecommunications in Europe a more extensive description of the technological evolution of the different phases is warranted. Unlike most other countries, Swedish Telecom decided to develop a fully automated system immediately. The Mobile telephone system A (MTA) was completed in 1952–3 and commercially launched in 1956. MTA worked in duplex (bi-directional traffic) with an automatic speech connection. Swedish Telecom did not actively market the service. It requested that it should be self-financing and at the same time prices should be low enough to attract at least high-paying customers. MTA remained a regional system, in Stockholm, Gothenburg and Malmö, with some 110 users, and was phased out in 1969. MTA suffered some substantial shortcomings: the telephones were unwieldy (40 kg), the service was only regional, the connection times were long and the system was difficult to use. From the early 1950s improvements were studied which determined the new MTB. Commercial service started in 1965 in Stockholm and Gothenburg, catering for 150 persons. In 1967–68 MTB was further extended, including Malmö, reaching some 500 subscribers. The system had an automatic speech connection and was based on the principle of dual tone, which meant that an exclusive selection tone identified the mobile telephone. The transition went to the fixed telecommunications network through the subscriber's relays with a unique subscriber card for each subscriber. This implied that the system could be used only if the subscriber had a subscriber card at several base stations. A time-out device was built into the system, a tone with increased intensity came on when conversations lasted longer than 3 minutes and continued until the connection was cut off. The weight of the subscriber unit was around 9 kg. MTB was dismantled in early 1983.

Neither MTA nor MTB generated any profit for Swedish Telecom, nevertheless, there seemed to be demand for this type of service. A report also recommended that the system should strive for nationwide coverage.

---

[9] See Mölleryd (1997) for an analysis of the evolution of the Swedish mobile telephone system.

It was considered extremely ambitious to have both an automatic and a nationwide system, but since also other Scandinavian countries were planning fully automated systems, the idea of a joint Scandinavian mobile telecommunications system was launched in 1969. Even though the long-term goal was the development of an automated system, the first assignment was a manual system, ready to be used immediately: it would take time to develop a fully automated system, and it was important to offer a mobile telephone service immediately. In 1971, the Scandinavian telecommunications conference approved the plans for a manual system, and decided on new rules which allowed the cross-border use of mobile telephones. The problem was that the computational requirements of the system for handling a large number of subscribers and base stations were simply too demanding to be done either manually or automatically with the computing power available at that time. The MTC system, which was supposed to address these issues, was never deployed as it coincided with the attempt of the Nordic countries to create a common system – MTC was the Swedish contribution to this. To cope with the large demand, a new MTD was introduced in 1971, also as an interim measure before cellular could be introduced. Operators from cord-operated switchboards at six service centres assisted subscribers, each operator filled in a form regarding the subscriber's number and length of the call. The system's radio parts were interconnected with the public telecommunications network at these service centres. The system had eighty channels and when fully extended 110 radio base stations. The system lay in the 460 MHz band. The system still lacked the possibilities for 'roaming'[10] and 'hand-over', however.[11] To place a call to a mobile telephone, the operator had to know roughly where the subscriber was located in order to direct the call over the nearest base station. It was an open system at first; the subscribers were called by their numbers, and everyone had to listen to the calling channel. This meant that other subscribers could also listen to calls in progress. When selective calls were introduced in 1974, no one had to wait for the calling channel but was instead given a signal. Concerning calls from a mobile telephone, the operator was attracted through tone signalling to activate the calling channel. The exchange indicated the relevant base station so that the operator could expedite the call.

The development of MTD started around Lake Mälaren and was gradually extended throughout Sweden. At its peak in 1981, the number of subscribers approached 20,000, and to relieve the bottlenecks handset

[10] 'Roaming' occurs when the subscriber of one network of one firm uses the network of another firm to phone.
[11] 'Hand-over' occurs when a mobile phone user moves from one cell to another without interrupting the phone call.

subsidies were introduced to induce customers to switch to the cellular system. MTD was phased out in 1987. The evolution of the Swedish system clearly shows the development of capacity: the MTA system had 141 subscribers at its peak in 1962, the MTB 659 subscribers in 1971 and MTD over 20 000 in 1981 in Sweden; it was also available in Denmark and Norway.

Germany In Germany, several mobile telephone network 'islands' emerged scattered across the country. The post and telecommunications operator Deutsche Bundespost merged them into the A-Netz[12] in 1958 (Jung and Warnecke, 1998). The interconnection between the mobile system and the public fixed telecommunications network was manually operated. After ten years of operation, the A-Netz covered about 80 per cent of the former Federal Republic of Germany and at its peak (1971) had 10–800 subscribers. The A-Netz was closed down in 1977.

Meanwhile, with the setting up of the B-Netz[13] in 1972, manual switching was replaced by automatic switching, the German territory was divided into mobile telecommunications areas and each had a prefix. To call a mobile subscriber from the fixed telecommunications network, it was necessary to dial the regional code and hence to know the region in which the mobile operator was located in that moment in order to establish automatic switching. In 1979 the B-Netz had reached its full capacity, with 13,000 subscribers, and covered the whole territory of the former Federal Republic of Germany. In 1980 the B-Netz also took over the frequency bands of the former A-Netz and could therefore expand its subscriber base to 27,000 in 1986. With a transmission power of 20 W, the base stations were able to cover an area of 25 km. Car phones had a transmission power of 10 W. The B-Netz could also be used by German mobile subscribers when 'roaming' in Austria, Luxembourg and the Netherlands. The B-Netz was closed down at the end of 1994.

UK The first British mobile telephone system (System 1) was introduced in 1959, but in a peripheral area in South Lancashire for testing purposes by the British Post Office (Garrard, 1998). The results were not very encouraging, and therefore deployment to London was not made before 1965. As for most of the telecommunications companies, this system was also not very profitable, even though used only by a restricted number of resourceful individuals. Capacity problems did not allow it to develop any further. System 2, the follow-up system designed by the

---

[12] This system operated in the 156–174 MHz frequency band and used a base station with 10 W transmission power.
[13] The B-Netz operated in the 146–156 MHz frequency band.

Post Office, was never deployed. Only with System 3, introduced in 1972, was the development of mobile telephones taken further. The system operated at 163 MHz, which implied a relatively small capacity as frequencies could not be reused within a range of 160 km. System 3 was still operator connected, and retained a press-to-talk switch on the handset This meant that only one person could talk at any time during a conversation. However, demand outstripped capacity and a waiting list was introduced in 1980.

In 1981 System 4 introduced a fully automatic direct-dial system and full duplex operation. The number of users peaked at about 14 000 in 1985, just when cellular systems were about to be introduced. The UK thus tagged behind with other countries such as Sweden and USA, both technologically and in terms of market development.

Other European countries Many other European countries introduced basic mobile telephone systems during the 1960s and 1970s. The systems typically were developed by PTTs in conjunction with their favoured national suppliers, and terminals were universally very expensive, a factor that limited demand to match the inherent low capacity. Although most of the systems were designed independently, they had many characteristics in common. The typical frequency range was 150–170 MHz. At this frequency, capacity was limited because of the small number of channels that could be used and the limited scope for reusing frequencies. Most of the systems required operators to connect callers, some even worked only in the press-to-talk mode. Technological innovation was thus needed to reduce cost and increase performance and capacity.

A major handicap for all these early mobile systems was that they required quite bulky and heavy user equipment. This implied that mobile phones had to be fitted as car phones although users calling from mobile vehicles could be interconnected into the public network. The modes of accessing the public network were different from country to country. Some required manual interconnection, whereas some, such as Sweden, developed fully automated switching right from the start.[14] One can conclude that radio communications was not a technology-led industry, but rather the opposite, well-identified applications had to wait until technology could satisfy them. Mobile communications is an example of how an application had to wait several decades until transistors were readily available before it was feasible for more than a few applications. Moreover, the applications of consumer electronics manufacturing techniques to mobile terminals brought prices down to a level that could be accepted by the mass market.

[14] For a detailed description see ITC (1993) and Mölleryd (1997).

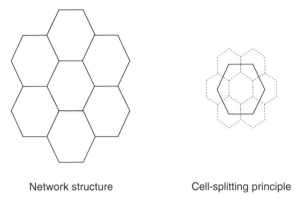

Network structure   Cell-splitting principle

Figure 2.1 *The basic working principle of a cellular network*
Note: The cellular mobile network may be represented as a web of cells. A frequency channel
is allocated to each cell which is different from that of adjacent cells. To increase subscriber
handling capacity, each cell can be split into smaller subcells and frequency channels are
reattributed accordingly.

2.2.2 The cellular concept

In the face of these capacity problems, it became clear that a more efficient
use of spectrum to support more subscribers and services requested an
entirely new system. The first ideas about 'cellular' networks had been
developed in the Bell Laboratories in 1947,[15] but its actual use had to await
the 1980s. Unlike the traditional approach to mobile telecommunications
but similar to radio or television broadcasting, the cellular system is based
on low-power transmitters, but lots of them, operating in specifically
designed smaller areas called *cells*. This may be sketched as follows.
Suppose a carpet of hexagons laid closely to each other, as indicated on
the left of figure 2.1. One hexagon is thus surrounded by six other hexagons. Each of these seven cells is served by a transmitter, called a base
station, and working at different frequencies. Frequencies used in a particular cell can be reused in non-adjacent cells for other users, as there is no
direct interference. This frequency reusage principle thus permits increasing capacity in proportion to the size of the cell. If an existing cell has

[15] The development was, however, left in an embryonic stage because the operation of
moving telecommunications units required an enormous amount of data processing.
Advances in microelectronics (the transistor, integrated circuits) made such tasks technologically feasible, but for a long time the costs remained prohibitively high. The large-scale
production and sharply declining prices of semiconductors such as microprocessors and
memories eventually made the cellular concept economically feasible by the 1970s. See
Calhoun (1988).

reached its capacity limits, it can be further subdivided into additional cells. This is referred to as 'cell-splitting', as indicated on the right of figure 2.1. Cell-splitting thus increases the scope for frequency reuse, and this permits an increase of traffic handling capacity, of course at the cost of an additional investment in base stations. Cell-splitting may be applied in a geographically selective manner: small cells are used for traffic-intensive urban areas and large cells are used in more suburban and less densely used areas. The cell size depends on a set of parameters, in particular the frequency used: the higher the frequency the smaller the cell.

There are four principles that characterise cellular mobile tele-communications:[16]
1. Lower-power transmitters and small coverage zones or cells
2. Frequency reuse
3. Cell-splitting to increase capacity
4. Hand-off and central control.

The cellular concept was developed to achieve a more efficient use of spectrum to support more subscribers. In contrast to the early systems, the cellular system makes use of low-power transmitters, operating at much higher frequency levels, typically in the range of 400–900 MHz. At these frequency levels, signals do not travel so far, so that the base stations have a limited reach and many base stations are required to obtain full coverage of a large desired geographic area. This implies a considerable investment. The crucial advantage is, however, that the frequency channels to support tele-phone conversations are locked only over a limited cell area: the frequency channels can be reused to support additional telephone conversations in other cells. A cellular system would not work with frequencies below 400 MHz, since signals would travel too far for reusing frequencies. As the frequency increases, the attenuation of the signals increases. This affects both the maximum and minimum feasible cell sizes. For example, a 450 MHz system is not suitable for urban areas with intense traffic because the minimum cell radius cannot go below 2 km. Likewise, an 1800 MHz system is good for urban areas but economically not justified for rural areas with little traffic since the maximum cell size of an 1800 MHz system is about 7 km.

A mobile cellular telecommunications system has five main components:
• Radio base stations or air interface
• One or more switches to control them and route calls
• A subscriber database
• A telecommunications network that connects the base stations and switches with the public telecommunications network
• A mobile subscriber terminal.

[16] For more details, see the appendix.

Taken together, the base stations form the *radio system*. This is the most critical building block of a cellular system. On top of carrying traffic, the radio system must continuously monitor the position of the user to route the traffic to the base station within whose range the user is located. As a user crosses a cell boundary, a new channel must be assigned quickly in order to maintain uninterrupted communication. This requires dedicated equipment, able to process large amounts of data.

The coordination of the communication activity within each cell, and possibly the control of a mobile terminal moving across cells, are daunting tasks in terms of data processing, which was not available in the 1940s and 1950s. It was necessary to wait until the advances in electronics permitted one to build switches that had the sufficient capability to handle the computational tasks for cellular technology. Electromechanical switching technology was far too slow to enable the 'hand-over' of users moving between cells during a conversation. The technological advances in micro-electronics, in particular the refinement of semiconductor technologies during the 1960s and 1970s[17] created the base for building faster electronic switches and suitable mobile terminals. During this period radio frequency technology also developed sufficiently to enable economic use of the higher frequencies needed. The large-scale production and sharply declining prices of semiconductors such as microprocessors and memories eventually made the cellular concept economically feasible by the 1970s. However, one barrier to its introduction remained. The frequencies in the 400–900 MHz range needed to be cleared; the lower frequencies used at that time by the existing mobile telephone systems were too low for frequency reuse, the principle on which the cellular concept is based. Regulatory reform to remove the previous users (e.g. in broadcasting) from the 400–900 MHz range of the spectrum took several more years (Calhoun, 1988). The first licences to cellular mobile telecommunication operators were eventually granted only at the beginning of the 1980s.

Although the main technological breakthroughs in microelectronic technologies which were key to the cellular mobile telecommunications industry mainly occurred in the USA,[18] the first cellular systems were actually put in place elsewhere. Because of the regulatory delays in assigning frequencies, the US was relatively late in deploying cellular mobile telecommunications networks. The first deployment and launch of services occurred in Japan in 1979 and in the Scandinavian countries in 1981, while in the USA it took until 1983.

---

[17] For a detailed description of the evolution of the technologies see Cortada (1987) and Morris (1990).

[18] See for instance Malerba (1985) and Morris (1990).

2.3 Characteristics of alternative cellular systems

2.3.1 Classification of cellular systems

Over time, different systems for cellular mobile telecommunications have been developed. Several technical features have been used to classify these systems. One may distinguish between two types of technologies according to the way in which the signals are transmitted: *analogue* and *digital* technology. Analogue signals are radio waves that vary in frequency and amplitude, digital signals consist of a stream of discontinuous pulses that correspond to the digital bits used in computers. Digital signals are divided into packets that are transmitted simultaneously with packets from other conversations (called 'multiplexing'). This process leads to a significantly more efficient use of the spectrum, thereby improving spectrum capacity by a factor of three to six (Rappaport, 1996).

Digital technology not only greatly improves transmission capacity, it also has several other advantages. For instance, it protects transmission integrity because digital pulses are more easily regenerated by computers. Moreover, a high transmission integrity in turn allows cellular operators to offer an expanding array of new data services (e.g. short messaging services). Finally, digital technology ensures privacy because digital signals cannot be eavesdropped.

A second important way to distinguish cellular systems is by the so-called 'access mechanism'. To increase the efficiency in the usage of the radio spectrum, the spectrum is divided into frequency bands, referred to as *channels*, which are then attributed to the different users. According to the way channels are attibuted, three different mechanisms can be distinguished.[19]

- *Frequency division multiple access (FDMA)*   In the early days of mobile telecommunications technology each user was attributed a channel, about 25 kHz wide, when she wanted to make a call, and each channel could be attributed to only one user at a time. This access principle is the basis for analogue cellular mobile telecommunications technology. This method requires only a limited computing power and permits simple mobile terminals. The disadvantage is that it requires the same number of transceivers for the base stations as the maximum number of simultaneous communication sessions. This method does not use the frequencies in a particularly efficient way.
- *Time division multiple access (TDMA)*   With TDMA, the channels are wider and divided into time slots (e.g. eight for GSM). Each user therefore

---

[19] For more details, see Garg and Wilkes (1996).

uses a wider channel (200 kHz), but for only a fraction of the time. TDMA thus requires precise timing between the transmitter and the receiver so that each user transmits during the time allocated to her. The advantages of TDMA compared to FDMA are the more efficient use of the radio spectrum. However, TDMA requires more complex mobile terminals and this technology is used in most of the current digital technologies.

- *Code division multiple access (CDMA)*   With CDMA, all communication sessions take place simultaneously in the available (relatively broad) frequency area. The result is that all users interfere. However, each user has a dedicated code that helps to identify the signal allocated to her from the 'noise' constituted by the other users. CDMA does not have a clearly fixed number of users as no exclusive capacity is allocated to a session. However, the quality of sound decreases with the total number of simultaneous users as the number of transmission errors increases. Whereas with TDMA each user of a transmission channel is allocated a time slot, with CDMA all users share the whole channel, but their signal carries a code to distinguish them from each other. To use an analogy: TDMA is like everybody speaking sequentially one after another; with CDMA, everybody speaks at the same time but with a different voice pitch that can be unambiguously captured by the receiver. CDMA is perceived to have a technical superiority over TDMA for data transmission, CDMA sends coded signals on a broad band of frequencies, and uses handsets that each listen to just its own code. A unique characteristic of CDMA is the 'soft hand-off', which allows user handsets to communicate with several base stations at the same time. Frequencies can be shared by adjacent cells, thus making frequency planning less complicated and cheaper than that required by TDMA-based technologies

Cellular mobile systems may also be classified by generations, according to the transmission capacity of the system. Whereas all analogue systems are also first-generation (1G) systems, digital systems are divided into second-generation (2G) and third-generation (3G) systems. Data transmission rates of 2G systems used to be limited to low speeds (such as 9.6 kbit/s) which is sufficient for voice services. Enhanced features such as HSCSD (High-speed circuit switched data), GPRS (General packet radio service) and EDGE (Enhanced data GSM environment) eventually permitted higher data rates (possibly up to 100 kbit/s) for certain 2G technologies.[20] 3G systems are referred to as systems that achieve higher data rates up to 384 kbit/s. Table 2.1 summarises the main feature of the different mobile telecommunications systems.

---

[20] In the trade press these upgrades are often referred to as '2.5G' technologies.

Table 2.1 *Different generations of mobile telecommunications technologies: key features*

	1G	2G	3G
Transmission mode	Analogue	Digital	Digital
Application	Voice only	Voice and low-speed data	Voice and high-speed data
Access technology	FDMA	TDMA, CDMA	CDMA
Number of incompatible systems adopted	7	4	2
First adoption year	1979	1991	2003

Looking at the historical evolution of mobile telecommunications technologies, one can observe an interesting trend towards *standardisation*. Whereas analogue technologies were developed independently in several countries at the same time, with digital systems there is an increased effort to make systems compatible. With 1G cellular mobile telecommunications seven analogue systems found application worldwide, with 2G this was reduced to four different digital systems. Although for 3G cellular technology, being installed after 2003, a single worldwide standard may not emerge, there should be at least a single family of two or perhaps three compatible systems.[21] The technological characteristics of the various analogue and digital systems are described in more detail below.

2.3.2 First-generation (1G) systems

All 1G cellular systems were analogue systems. The large number of analogue systems in the early days of the cellular industry may be explained by the fact that most countries viewed cellular telecommunications as just an additional new business of the state-owned telecommunications monopoly. The development of the cellular network was thus a means of honing the innovative capabilities of national equipment suppliers. Some technological features of the various analogue systems are summarised in table 2.2. Among the most important differences between analogue systems are the frequency range allocated for transmission and

---

[21] The International Telecommunications Union (ITU) defined five different systems as technological options for 3G mobile telecommunications. Out of these, only two based on CDMA technology have found application with existing systems. For a discussion of the technological options, see Webb (1998) and Gruber and Hoenicke (1999).

Table 2.2 *Characteristics of 1G (analogue) cellular systems*

System	Year of first adoption	Country of first adoption	Transmission frequency (MHz)	Channel band width (kHz)	Number of speech channels	Channel bit rate (Kb/s)	Spectral efficiency (b/s/Hz)
NTT	1979	Japan	400–800	25	1640	0.3	0.012
NMT-450	1981	Scandinavia	450–470	25	180	1.2	0.048
NMT-900	1986	Scandinavia	890–960	12.5	2000	1.2	0.096
AMPS	1983	USA	824–845	30	832	10	0.333
C-450	1985	Germany	450–465	20	573	5.28	0.264
TACS	1985	UK	890–960	25	1000	8	0.320
RTMS[a]	1985	Italy	450–465	25	200		
RC 2000[a]	1985	France	200–400	12.5	1700		

Notes: [a]Also referred to as 'quasi-cellular', because of restrictions on handover between cells.
1 MHz = 1 million Hertz; 1 kHz = 1000 Hertz; b = bit; s = second
Source: Author, based on Garg and Wilkes (1996) and ITC (1993).

the bandwidth of a channel. The frequency range and the channel band-width determine the number of speech channels. The channel bit rate (in 1000 bits per second) indicates the density of the bit-stream. The spectral efficiency is the number of bits that can be sent per second over a channel of a given bandwidth – i.e. the channel bit rate divided by channel bandwidth. This ratio may be used as a very rough measure of the efficiency of a system.[22]

The individual systems can be characterised as follows:

- *NTT*: Japan introduced this as the first cellular system worldwide in 1979. A comparison with the other 1G systems indicated in table 2.1 shows that this is a rather inefficient system in terms of channel bit rate at a given bandwidth. This system did not find application outside Japan.

- *NMT* (Nordic mobile telephony system): This is the cellular system introduced in 1981. It was jointly developed as NMT-450 by the Scandinavian countries Denmark, Finland, Sweden and Norway, and then adopted as the single cellular standard for these countries. Around 1983 there were already signs of capacity shortages and additional spectrum was required, and so the NMT-900 was introduced in 1986. Since the NMT-900 system had been assigned more spectrum and needed a smaller channel width, the number of channels could be increased from 180 to 2000. Furthermore, the system was specified as a small-cell cellular system whereby cells in each transmission area could be split further into so called 'micro-cells'. The ensuing lower power requirements enabled the construction of smaller handsets, as signals had to travel a shorter distance. From the point of view of the equipment producers, the NMT-900 specifications did include innovative features, but these were mostly related to the software requirements of the equipment that had already been developed for the NMT-450 system.

- *AMPS* (Advanced module phone service): The principles of this system was proposed by the US telecommunications firm AT&T in 1970, which were at that time part of the Bell group which had a monopoly franchise for the provision of telecommunications services and equipment. It took the FCC a further twelve years before the decision to license the system was taken. During these years there was a heavy regulatory debate concerning the breaking up of the Bell system, the number of licences to be allocated in each geographical area and to what extent AT&T could be supplier of both equipment and telecommunications services.

- *TACS* (Total access communications system): This system, introduced in 1985, was a British adaptation of the AMPS standard to comply with

---

[22] For more details see Mehrotra (1994).

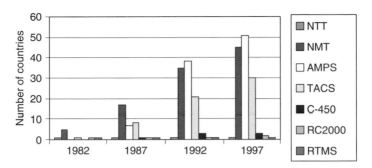

Figure 2.2 *Diffusion of 1G technologies, by number of adopting countries, 1982–1997*
Source: Gruber and Verboven (2001a).

the different frequency allocation prevailing in Europe.[23] The two systems are not compatible, but many specifications are similar and thus the equipment suppliers could use the same network specification.

- *C-450*: This system was first adopted in Germany after the failure to reach an agreement on a joint procurement with France.[24] The system was developed by Siemens, the dominant national telecommunications equipment supplier in Germany. It was technically complex and already incorporated several features that would be taken up later by GSM, such as security (thanks to a subscriber identification module, SIM). A general approach of the system was that it apparently showed little regard for user needs. C-450 was adopted only in Portugal and South Africa, consequently, it could not attain the benefits from economies of scale that the systems in the USA, the UK and the Scandinavian countries enjoyed.

- *RC 2000 and Radio telephone mobile System(RTMS)*: These systems, adopted by France and Italy respectively, are sometimes referred to as 'quasi-cellular services', because there was limited ability for hand-over from one cell to another. These systems were soon complemented or replaced by other analogue systems.[25]

Figure 2.2 illustrates the differences in the worldwide diffusion of the various analogue systems. During the early years of the cellular industry, NMT was the most widespread system, reflecting the early lead in the adoption of cellular telecommunications in the Scandinavian countries. However, at the end of the 1980s AMPS had the largest number of national

---

[23] See Appleby (1991) for a description of this adaptation.

[24] See Müller and Toker (1994) for description of these joint efforts.

[25] Both countries later adopted additional systems: the NMT system in France (Manguian, 1993) and the TACS system in Italy (Guerci *et al.*, 1998).

networks. On the basis of spectral efficiency, this system also ranked first (table 2.2). NMT, which was actually one of the least efficient systems, nevertheless kept the largest number of networks in Europe and ranked second worldwide. The TACS system ranked third in terms of number of networks. The C-450 system, which was technically quite sophisticated and efficient, was adopted in only two countries besides Germany. RC 2000 and RTMS found no adoption outside France and Italy.

This comparison suggests that the most successful systems occur if the domestic market is sufficiently large (AMPS for the USA) or if the government coordinates with other countries (NMT for the Scandinavian countries). The large diffusion of TACS follows from the fact that it is an adaptation of AMPS to another frequency band. The examples of NTT, C-450, RTMS and RC 2000 show that the national market alone is usually too small economically to support the development of additional incompatible systems.

2.3.3 Second-generation (2G) systems

The number of digital cellular systems is much lower than the number of analogue systems, mainly because the European countries this time cooperated over the development for a GSM standard. Backward compatibility with existing analogue systems was not a concern, since it could hardly have been resolved in any case given the large number of analogue systems. A 'clean sheet' approach was thus adopted in the design of the new system. The USA adopted a different strategy. In contrast to the 1G system where a national standard was mandated, for 2G this was no longer the case. Firms were left free to adopt the appropriate technology, provided that there was backward compatibility with the existing system. Japan, for example, its own 2G system. Table 2.3 provides a summary of some technological features of the various second generation systems.

There are two large families of 2G cellular systems, characterised by their so-called 'access systems': TDMA and CDMA. The TDMA technology splits a frequency channel into n different time slots, and allocates each user one time slot. Thus n calls can travel over one cellular channel, compared to only one under an analogue system. The three incompatible TDMA systems that found application are: GSM, D-AMPS and PDC. CDMA allows all users to share the whole frequency channel, but the signals carry a code to distinguish them from each other. The advantage of CDMA is that the same set of frequencies can be used in every cell, due to a peculiar ability to discern signals from noise. This provides potentially great improvements in capacity. The only 2G system using the CDMA

Table 2.3 *Characteristics of 2G cellular systems*

System	Year of first adoption	Country of first adoption	Access technology	Transmission frequency (MHz)	Channel band width (kHz)	No. of speech channels	Channel bit rate (Kb/s)	Spectral efficiency (b/s/Hz)
GSM 900	1990	EU	TDMA	890–960	200	1000	270.8	1.35
GSM 1800	1993	EU	TDMA	1710–1880	200	1500	270.8	1.35
D-AMPS	1991	US	TDMA	824–894	30	1666	48.6	1.62
IS-95	1993	US	CDMA	824–894	1250			1.75[a]
JDC	1993	Japan	TDMA	800–1500	25	1920	14	1.68

Note: [a] Not strictly comparable; value based on several restrictive assumptions.
Source: Author, based on Garg and Wilkes (1996) and Webb (1998).

technology is IS-95. In the following the four 2G systems are now described briefly,

- *GSM*: The acronym GSM stands for Global system for mobile communications.[26] This system was developed as a coordinated effort by the European countries during the second half of the 1980s. The GSM standard has a spectral efficiency that is about four times higher than the most efficient analogue system, but among digital systems it is the least efficient.[27] Initially, it was developed for the 900 MHz frequency range, but during the second half of the 1990s it was adopted also for the 1800 MHz band. Thanks to 'dual band' handsets, mobile telecommunications firms could use the two frequency bands indiscriminately.

- *JDC*: The system, which has similarities with D-AMPS below,[28] was introduced as a national standard in Japan. The promoters of this system made attempts to spread it in the Pacific region under the name Pacific Digital Cellular (PDC). This, however, had little success as no other country outside Japan adopted the system.

- *D-AMPS*: This system was first introduced by the USA under the name IS-54. The objective was to ensure a smooth transition of the prevailing analogue standard into a digital system. The system divides an analogue channel into three parts, thereby tripling capacity. It turned out that IS-54 had a worse speech quality in the digital mode compared to the analogue mode, this was, however, improved with a later revision of the system, now also known as IS-136. D-AMPS has a higher spectral efficiency than GSM.

- *IS-95*: This system is based on the innovative CDMA technology, developed and patented by the USA firm Qualcomm under the name cdmaOne. CDMA technology is an outgrowth of defence applications. Qualcomm had a very aggressive strategy of announcement of technology features, granting the technology a much broader audience than the initial performance of the technology actually warranted.[29] For instance, it was claimed that the capacity advantage over GSM was a factor of more than 20. This is certainly exaggerated, though a widely shared guess of the capacity advantage of CDMA over TDMA is about 30 per cent. (Garg and Wilkes, 1996; Webb, 1998).

Figure 2.3 shows the differences in popularity between the various digital systems. GSM was the first to be introduced in a large number of countries and since then it has remained by far the most widespread system in

---

[26] Initially it actually was for *Groupe système mobile*. For an institutional history of GSM, see Garrard (1998) and for a detailed technical description Redl, Weber and Oliphant (1995).
[27] This statement, however needs to be qualified as there are also other parameters that need to be taken into consideration. For details, see Mehrotra (1994).
[28] See Mehrotra (1994) for a discussion of affinities.
[29] Steinbock (2003) describes in more detail the marketing strategy of Qualcomm for promoting this technology in the engineering community.

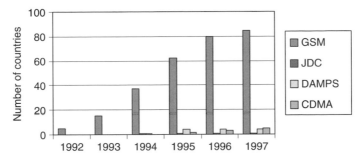

Figure 2.3 *Diffusion of 2G technologies, by number of adopting countries, 1992–1997*
Source: Gruber and Verboven (2001a).

terms of both adopting countries and subscribers. In 1997, of the 40 million digital subscribers worldwide, more than 80 per cent were GSM subscribers (ITU, 1999). Of the competing systems, D-AMPS had a timing advantage over CDMAIS-95, as well as the backward-compatible installed base of AMPS. However, cdmaOne is spreading more rapidly, especially in Asia, where it has won the race against JDC. Moreover CDMA 2000, the 3G system based on CDMA, has the advantage of having the same technology base as cdmaOne, which gives it the advantage in the USA of permitting existing 2G firms to supply 3G services without requiring additional spectrum.

2.3.4 Third-generation (3G) systems

Whereas the first and second generation of mobile telecommunications systems were mainly designed for voice transmission, the next technological step was the development of systems for *data transmission*. 3G systems are designed significantly to increase data transmission rates. ITU promoted a global standard for 3G mobile telecommunications through the initiative IMT-2000, characterised by the following features: first, seamless global 'roaming', enabling users to move across borders using the same number and handset; second, an at least 40 times higher signal transmission rate than 2G systems, allowing for fast Internet access.

The ITU has accepted five systems for the family of IMT-2000 standards that satisfy technical requirements to provide 3G services. Three of them are based on CDMA and two on TDMA. Only those based on CDMA are expected to find widespread adoption. They are:

- W-CDMA, wideband CDMA, known also as UMTS and promoted by ETSI (European Telecommunications Standardisation Institute) the European standards body

- CDMA 2000, a further development of the existing cdmaOne technology promoted by the CDMA patent holder Qualcomm
- IWC, a system based on TDMA (compatible with the US system D-AMPS as well as GSM) and promoted by China under the heading TD-SCDMA.

The failure to reach an agreement on a 3G standard arises from the need for multiple-mode and probably also multiple-band handsets capable of handling various modes and frequency bands. This would enable worldwide 'roaming' but at a higher cost than with a single standard because of the increased complexity of handsets and networks.

The failure to agree on a world standard is also due to the fact that mobile firms seek backward compatibility for their installed mobile systems. In Europe 3G systems are referred to as UMTS, a concept developed by the ETSI. A European Directive instructed member states to assign licences for 3G mobile telecommunications services and stated that at least one of the licence holders should adopt W-CDMA as its technology. (W-CDMA has the advantage that it is backward compatible with GSM.) Clearly, the European interest was in making UMTS backward-compatible as much as possible given the large installed base. It is, however, very likely that the large majority of the 3G service providers will also adopt W-CDMA. This objective conflicts with making it compatible with CDMA 2000, the further development of the existing IS-95 technology, which relies on CDMA and has a large installed base in the USA.

The first adoptions of 3G systems started in 2002 in Japan and in 2003 in Europe. The USA delayed the introduction of 3G systems, mainly because of the slow development of 2G systems which were launched late and used a range of different, non-compatible 2G technologies. European policy makers were very keen to introduce early 3G systems, since early adoption of UMTS was seen as key for preserving the worldwide lead in mobile telecommunications technologies established with GSM.[30] In the aftermath of the stock market 'bubble' bursting in 2000 and the ensuing difficulties in finding financial resources for the construction of 3G mobile telecommunications networks, questions about the profitability of the investment in 3G mobile telecommunications have begun to arise. This essentially boils down to the question of whether the speed of adoption proposed and the size of required investments are warranted by a sufficiently high level of demand for 3G services. Several simulation exercises[31] have shown that revenues from data services will have to increase

---

[30] See European Commission (1997a, 1997b) on the statements for industrial policies in the mobile telecommunications sector.
[31] See for instance, Gruber and Hoenicke (2000), Didier and Lorenzi (2002) and Björkdahl and Bohlin (2003).

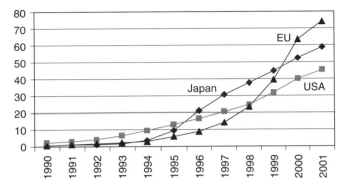

Figure 2.4 *Penetration rate of mobile telecommunications in the main developed country regions, 1990–2001 (subscribers/100 inhabitants)*
Source: OECD data.

substantially to make UMTS a profitable undertaking, but firms will be exposed to substantial high risks if they introduce the new technology too early.

2.4 Subscriber trends

2.4.1 Penetration rate

During the first years of the cellular mobile telecommunications industry, growth in terms of subscribers was modest in most countries, with a tiny fraction of the population subscribing. As we have seen, the most advanced countries for mobile telecommunications were the Scandinavian countries and the USA. It was only during the 1990s that the market for mobile telecommunications really started to take off in other countries. Figure 2.4 shows the historical evolution of the penetration rate for mobile telecommunications for the EU, the USA and Japan. The time path is an interesting sequence of leapfrogging movements among the various geographical regions. The USA was leading in terms of penetration rate until the mid-1990s, i.e. during the whole period of analogue technology, it was then overtaken by Japan, which had a short spell of leadership. The penetration rate in the EU expanded very rapidly during the second half of the 1990s and had finally overtaken the USA by the end of the 1990s.

Such growth rates in subscribers could be observed across all the regions in the world, both developed and developing countries. In the OECD area for example, by the end of 2001 the number of cellular mobile subscribers reached 612 million, which is about five times more than there were in 1996

Table 2.4 *Evolution of mobile and fixed telecommunication subscribers, 1996–2001*

	Lines (million)		Annual growth (per cent)	Penetration rates[a]	
	1996	2001	1996–2001	1996	2001
Mobile					
OECD	120	612	39	11.0	53.9
Non-OECD	24	328	69	0.5	6.5
World	144	940	46	2.5	15.3
Fixed					
OECD	500	517	1	45.8	45.5
Non-OECD	230	518	18	5.0	10.3
World	730	1035	7	12.7	16.8

Note: [a]Penetration rate is the number of mobile subscribers per 100 inhabitants.
Source: ITU data.

(see table 2.4). For non-OECD countries, the number of subscribers in 2001 was 328 million, which was over thirteen times more than in 1996. Overall, the number of mobile subscribers worldwide grew at a compound annual growth rate of 46 per cent between 1996 and 2001, implying that subscribers more than doubled every two years. Nevertheless, there are still large differences in penetration rates across various countries, even within country groups with the same level of income.[32]

Table 2.4 also lists the evolution of fixed lines. It shows that in comparison with fixed line telecommunications, where the growth of lines is modest, mobile telecommunications networks are growing at a very high speed and the number of mobile telecommunications subscribers has in many countries overtaken the number of fixed lines.[33] In several countries, the number of fixed lines is actually falling, suggesting a substitution effect between fixed and mobile telecommunications. Recent economic literature has investigated the determinants of this rapid growth in demand for mobile telecommunications. There are an increasing number of studies that look at issues such as comparing differences in the evolution between industrialised countries, focusing on the role of country characteristics (Ahn and Lee, 1999), productivity effects for the telecommunications sector (Jha and Majumdar, 1999) or the importance of macroeconomic variables on the evolution of the industry (De Kimpe,

[32] See, for instance, OECD (2003) for detailed market data.
[33] ITU (2003a) reports that during 2002 the mobile subscribers overtook fixed line subscribers worldwide.

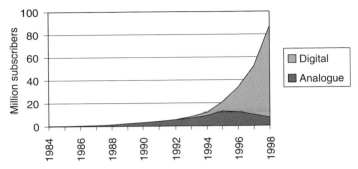

Figure 2.5 *Diffusion of analogue and digital mobile telecommunications, EU,*
1984–1998
Source: ITU data.

Parker and Sarvary, 1998). Several of these aspects will be dealt with in
more detail in chapter 4.

2.4.2 Growth drivers

In the early 1980s, most Western European countries introduced 1G
systems, mostly in monopoly regimes. In Europe, these networks were
based on a variety of incompatible technologies, so that international
'roaming'[34] was not possible. The subscriber capacity of these early networks was limited and mobile penetration of the population remained low.
In countries such as the UK and Sweden, where analogue mobile telecommunications were supplied by a duopoly, penetration rates were above
the European average. In the USA, analogue technology was supplied in a
duopoly market structure in regional markets. All analogue USA networks used AMPS technology which made USA-wide 'roaming' feasible.
Penetration rates in the USA were higher than in most European
countries.

In Europe, a major breakthrough in mobile communications occurred
during the mid-1990s with the switch to digital technology. The introduction of GSM was coupled with an appropriate and efficient regulatory
environment which facilitated the spread of the technology. Figure 2.5
illustrates the expansionary impact of digital technology; the growth in the
sector since 1995 coincided with the introduction of 2G technology while
the analogue subscriber base in the EU was shrinking. By the end of the

---

[34] 'International roaming' refers to the feature that a mobile user takes her handset abroad
and is able to make calls in the same way as at home.

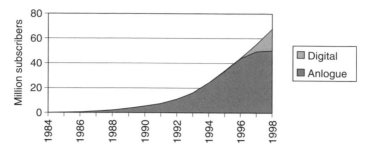

Figure 2.6 *Diffusion of analogue and digital mobile telecommunications, USA, 1984–1998*
Source: ITU data.

decade many of the analogue networks in the EU had been switched off, but some were kept alive.[35]

There is a widely held view that competition contributed to the rapid diffusion of mobile telecommunications in Europe.[36] The introduction of competition in the form of entry by new firms coincided in many European countries with the introduction of digital GSM networks. In fact, EU Directives required member states to grant at least two GSM licences for the 900 MHz frequency band and at least one further licence for the 1800 MHz frequency band. In an econometric study of the EU countries, Gruber and Verboven (2001a) showed, however, that the expansionary effect deriving from the switch to digital technology was much stronger than the effect deriving from introducing competition.

A comparison with the USA (figure 2.6) shows that there digital technology appeared to have a lesser impact on the diffusion of mobile telecommunications. A major reason was the absence of a uniform standard of digital mobile telecommunications technology: the USA has three incompatible digital technologies. This deprives users of several benefits with a national standard. These include nationwide 'roaming', the possibility of changing service provider without the need to change the handset, as well as a cheaper and wider choice of handsets.

These comparisons suggest two key elements that may have contributed to the success of the new generation of mobile technology in the EU:

---

[35] The largest analogue network in Europe was the TACS network of the Italian firm TIM, which at the beginning of 2000 still had more than 3 million subscribers.
[36] See, for instance, ITU (1999), OECD (1999).

- *Superior technology*: The digital approach led to a breakthrough in performance, capacity and quality. Size and cost of equipment, most importantly of mobile phones, could be dramatically reduced.
- *Standardisation*: Publicly available standards guarantee that equipment from different manufacturers is compatible and the user benefits from a broader choice at lower prices as producers have less scope for market segmentation.

As a result, powerful mobile network equipment and mobile phones became available at ever-decreasing prices and better quality. The switch to digital technology in mobile telecommunications permitted a more efficient usage of the radio spectrum and an increase in the number of subscribers. Thus operating firms could better exploit economies of scale, and greater capacity enabled competition between network operators. Economies of scale in the equipment manufacturing process and competition among equipment manufacturers continued to bring down equipment prices. The required investment per subscriber was therefore falling and customer segments with lower usage, which previously were unprofitable to serve, could now be targeted, too. This may help to explain why the mobile communications service has now become a mass market product.

2.5 Evolution of mobile telecommunications revenues

Telecommunications technology has a growing role in advanced industrialised economies. Table 2.5 shows that for the industrialised countries the share of the total telecommunications market as percentage of GDP has increased over time and represented 2.8 per cent of EU GDP in 2001; in

Table 2.5 *Weight of mobile telecommunications in the telecommunications sector and the economy, 1993–2001*

	Percentage share of total telecommunications in GDP			Percentage share of mobile telecommunications in total telecommunications		
	1993	1997	2001	1993	1997	2001
EU	2.0	2.3	2.8	4.5	15.7	31.1
USA	2.8	3.2	3.5	5.9	12.9	22.2
Japan	1.6	3.0	3.4	11.5	39.7	53.1

Source: OECD data.

the USA and Japan it was even higher, representing, respectively, 3.5 and 3.4 per cent of GDP. Total revenues for mobile telecommunications are increasing faster than revenues for fixed telecommunications, thanks to the strong growth in subscribers. Thus the share of mobile telecommunications revenues in the total telecommunications sector is growing over time as well, although there may be notable differences across countries. Mobile telecommunications accounted for 22.2 per cent of total telecommunications revenues in the USA in 2001, for the EU, the share was considerably higher, at 31.1 per cent, reflecting the much higher subscriber base achieved, especially during the second half of the 1990s. For Japan, the share of mobile telecommunications in the total telecommunications sector is even higher, at 53.1 per cent, fast overtaking the fixed line sector.[37]

The revenues for mobile services are for a large part generated by *traffic* and *tariffs*. While traffic revenues per minute for mobile traffic tend to be higher than for fixed line services,[38] mobile calls tend to be shorter than fixed line calls. As the number of subscribers grows, there are an increasingly larger number of mobile subscribers that generate low traffic. In spite of the fact that mobile traffic has been growing at a very fast rate, overall traffic originated by mobile networks is smaller than the traffic generated by fixed networks.

2.5.1 Usage patterns and average revenue per user

One can apply the product life cycle model[39] to the evolution of the mobile telecommunications market with the mobile phone market going through the familiar features of start-up, expansion and maturity (see figure 2.7). The start-up phase of mobile communications markets has been characterised by limited or little competition. Monopolies or duopolies penetrated the segment of high-spending, price-insensitive users, as indicated by the average revenue per user (ARPU). Typically high ARPUs are observed with early adopters, once a high penetration of this first-segment has been achieved, and often in anticipation of a further entrant to the market, the monopoly operator or duopoly operators starts penetrating more price-sensitive segments, such as small and medium enterprises (SMEs). The market enters the expansion phase where growth in subscribers is strong. While it is to be noted that there can be serious competition in a duopoly, usually a third

---

[37] ITU (1999) indicates that in the world market for telecommunications in 1998, mobile telecommunications accounted for 21.2 per cent. The ITC also estimated that the mobile sector would overtake the fixed line sector in terms of revenues in 2004 (ITU, 2003a).

[38] For details see chapter 5.

[39] For a survey see Mahajan, Muller and Bass (1993).

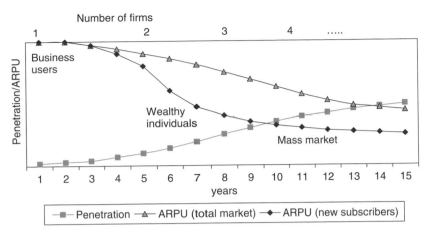

Figure 2.7 *Stylised representation of the evolution of the penetration rate and average revenue per user (ARPU)*

entrant is required to start competition that addresses the true mass-market segment, including low-spending users as well.

In the mobile telecommunications industry, technological progress is rapid and readily available to all operators in the market, so that innovative services can easily and quickly be copied by the competition. As a result, telecommunications services soon became a homogeneous good with little scope for differentiation. Competition focuses therefore mainly on price. Increased competition leads to price cuts and makes mobile services affordable for the low-spending consumer segment, or mass market. While the market continues to expand in terms of subscriber numbers, falling tariffs and an increasing portion of lower usage customers counterbalances the subscriber growth to the extent that growth in revenues slows down, halts or even reverses, at least temporarily.

The market reaches maturity when no additional subscribers can be added to the market. The market is then *saturated*. Once this stage is reached, price competition will be most intense, as operators need to attract the competitors' subscribers. ARPU per subscriber is likely to stabilise. In practice, two opposing forces are determining the usage pattern of the average mobile telecommunications subscriber. First, as the penetration rate of mobile telecommunications increases, more and more low-usage subscribers enter the market, the average usage or traffic per subscriber declines. Second, for existing subscribers declining tariffs may induce an increase of usage.

Thus, as long as the effect deriving from the addition of new low-usage customers prevails, we have an overall decrease in traffic per subscriber.

Overall traffic per subscriber should stabilise or even increase once the second effect predominates – i.e. the addition of new subscribers is relatively small. Empirically, it is most likely that the second effect will prevail.[40]

However, even if traffic per subscriber increases, revenues need not, if tariff declines more than compensate traffic increases. ARPU is one of the most important business parameters for mobile phone operators and a benchmark for the profitability of a firm. ARPUs used to be very high in mobile communications, especially if compared with fixed line communications. The reason was that in the early phase of the industry the typical subscriber was the business user, who had a high usage and low price elasticity. ARPU declines as the penetration rate increases and low-usage subscribers are attracted by low tariffs. As figure 2.8 shows, there is typically a negative correlation between ARPU and penetration rate in the context of a cross-country comparison.

This negative correlation is even more evident in the time series dimension within a country. Figure 2.9 indicates the relationship between ARPU and penetration rates over time for the Finish telecommunications firm Sonera (formerly Telecom Finland). As the penetration rate steadily increases in Finland, Sonera's ARPU declines. Considering the already high penetration rate Finland has achieved, the absolute level of ARPU is relatively high. This picture is representative of what is happening to the firms in the industry, in fact, for some firms, the ARPU has a lower intercept or is declining more rapidly.

Mobile telecommunications involve a *two-way network*: calls initiated by a subscriber of a certain network may be terminated on a different network and, conversely, a certain network will terminate calls originated on other networks. Apart from the outgoing traffic generated by its own customers, traffic termination is a major revenue item for a mobile telecommunications firm.[41] A firm charges for terminating the incoming traffic attracted by its customers. Once a user has decided to join a particular mobile firm, that firm has a monopoly position over termination services to that subscriber, as reflected in unusually high termination charges applied by mobile firms. Mobile telecommunications firms traditionally received much more for terminating a call on their network if the call was originated from a fixed network compared to the reverse direction. However, the cost of conveying a particular call from a point of interconnection to its destination on the terminating fixed network is basically the

---

[40] For instance, in Finland monthly outgoing minutes per average subscriber moved from 90 in 1995 to 97 in 1997, an average annual increase of 4 per cent (Ministry of Transport and Communications Finland, 1998).

[41] This may not apply for countries where charging is based on RPP; for details, see chapter 5.

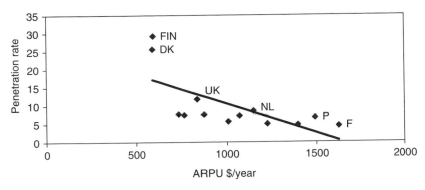

Figure 2.8 *ARPU and penetration rate (mobile users/100 population), EU countries, 1996*
Source: Author's calculations based on company accounts.

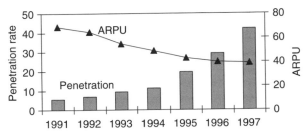

Figure 2.9 *Typical evolution of penetration rate and ARPU, 1991–1997*
Note: The average revenue per user (ARPU) is in US dollars per month for the Finnish firm Sonera. The penetration rate refers to the number of mobile telecommunications subscribers per 100 inhabitants in Finland.
Source: Sonera data.

same whether the call originates on a mobile network or another fixed network. There is thus no justification for the large differences in inter-connection charges imposed by mobile firms depending on the type of network on which the call originated.[42] Interconnection rates vary greatly across Europe. In 1997, for instance, Ireland had a 600 per cent higher termination charge above what was considered 'best practice' (European

---

[42] For instance, in Europe the relevant Directives require firms to proceed on a non-discriminatory basis between fixed and mobile operators when establishing interconnection tariffs. Ideally, interconnection rates should be established with reference to the long-run incremental cost. Most regulators request that operators implement this principle, but difficulties arise in establishing a common view how the long-run incremental cost should be calculated.

Commission, 1997a, 1997b). Although the European Commission has been quite successful in reducing these discrepancies, they still remain significant: in 2001, there still was a 250 per cent difference between the lowest and the highest termination charge within the EU (European Commission, 2002c). These high termination prices were initially justified by the mobile telecommunications firms through the higher set-up costs of their new mobile networks compared to the fixed line networks which were supposedly depreciated to a large extent. These arguments were, however, increasingly contradicted by reported evidence on the substantial cross-subsidies from the fixed to the mobile sector, which again led to strongly distorting pricing practices at the retail level. High termination charges can be used as enforcement devices for collusion and the proceeds used for subsidising subscriber acquisition. As a result, both regulatory and competition authorities began to become interested. This will be discussed in more detail in chapter 5.

A further source for revenues are 'roaming' services. International 'roaming' raises similar issues as termination charges: when firms are competing for the same market there is in principle scope for relaxing price competition by building networks of differing coverage (see Valletti, 2003). This can happen, in particular, in the early stages of entry by new firms. To sustain such a strategy, 'roaming' agreements within a country should not be permitted,[43] because firms would otherwise lose their differentiation capability and would price too aggressively. On the other hand, firms should seek 'roaming' agreements with foreign operators, since this would bring a beneficial market expansion effect and at the same time they would not be competing for the same customers. International 'roaming' agreements are in fact common practice in the mobile industry (as long as operators in different countries use compatible standards). They are typically on a voluntary base since there is a 'double coincidence of wants' for the two parties. It should be noted that this latter remark is valid as long as operators stay in different countries; the exception is when firms do not have nationwide licences. They then have incentives to make voluntary 'roaming' agreements with firms that have complementary coverage. This happened in the USA, where there were no nationwide licences and firms had to make 'roaming' agreements to increase coverage beyond the respective licence areas. This was the alternative to merging networks, though early national 'roaming' was cumbersome from a technical point of view.[44] The US wireless industry saw a flurry of mergers until at the end of the 1990s five nationwide cellular firms had emerged.

---

[43] Regulators, however, typically enforce national 'roaming' when incumbents have to help new entrants.
[44] Initially 'roaming' was of the manual type: the user had to register and provide payment credentials outside her home area before she was able to make any calls. Likewise calling parties had to know in which area the called party was 'roaming'. This contrasts with

International 'roaming' became important only once internationally compatible systems, such as GSM in Europe were put in place.[45] International 'roaming' is of the automatic type – i.e. users do not have to make separate arrangements with the hosting firms. The wave of mergers and acquisitions (M&As) at the international level that occurred at the end of the 1990s may change the incentives to provide international 'roaming': if global operators face each other in different countries, they may become more reluctant to grant 'roaming' to rivals. Denying 'roaming' (or giving it only at an exceptionally high charge) is a cost-raising strategy that would not allow a rival operator to compete effectively for those corporate customers that value coverage highly. An operator may thus find it necessary to bypass international 'roaming' by either investing in the foreign country or by merging with or buying a foreign operator.

2.5.2 Trends in product differentiation and pricing

The growth rates in mobile users reflect some typical trends in the pricing of services, which are related to the underlying market structure. Entry and more intense competition resulted in the innovation of flexible tariff packages targeted at different categories of users rather than price cuts.[46] As will be seen, innovative pricing strategies were responsible for the first wave of growth almost everywhere, made possible thanks to the move from analogue to digital technologies that brought with it additional capacity.

A precise definition of the market for mobile telecommunications services is difficult, as there are several ways to define the services. One can make a basic distinction between *wholesale* services and *retail* services. Because of the spectrum constraint there is a very small number of wholesale firms – i.e. firms that can set up and operate a mobile telecommunications network. However, there is no reason to limit the number of retail firms (or service providers) which buy bulk services from the wholesaler

automatic 'roaming', where all the steps are done automatically. In the USA the practice of manual 'roaming' was widespread, while manual 'roaming' has now become mandatory, the FCC has also discussed whether automatic 'roaming' should become so.

[45] International 'roaming' was cumbersome in the analogue phase because of the large number of incompatible systems; it was feasible only in the Nordic countries, which had a common NMT system.

[46] On the basis of per-minute costs, mobile telephony is still quite an expensive service. Fixed network operators have responded to increased competition from falling mobile telecommunications service prices by cutting the prices for fixed line calls. In the Scandinavian countries, for instance, where the mobile telecommunications sector is most advanced, mobile services at the end of the 1990s were on average still at least five times more expensive than fixed telecommunications services, if calculated on the per-minute price of a call (OECD, 2000).

and repackage them for retail sale to the end user. The end user makes decisions about which handset to buy, what tariff bundle to sign up to and the range of services to use. The necessary services are registration and access to the network, i.e. call origination and termination. Optional services include items such as provision of handsets, international 'roaming' and messaging services. Firms can use these individual service items strategically to segment the market and to differentiate themselves from competitors. The main means for product differentiation is in the *bundling* of the service, targeting the usage behaviour of different customer profiles. In the early period of the market (roughly corresponding to the 1980s), there was little scope for the product differentiation. Mobile markets were led by business demand and users were prepared to pay high prices for the service. Leading firms operating in monopolies or duopolies did not adopt new strategies for the expansion of personal communications until they faced increasingly competitive markets. Pricing was undertaken on a uniform basis with no variation made for users with contrasted usage patterns.[47] Pricing strategies were simple and mostly designed to ration the available spectrum capacity – for instance, by charging higher prices in densely populated areas.

In the stage of take-up (early 1990s), operators started to address new types of users (mobile professionals, the self-employed, salespeople) with a wide range of tariff packages. Most market structures were duopolies, demand was growing and there was no big incentive for mobile operators to cut prices, rather they tended to differentiate from fixed line prices (typically, mobile telephony prices are not distance-sensitive). The entry of new operators forced incumbents to be more responsive and growth was driven more by product differentiation than by price reductions.[48]

The process of price differentiation continued in the mid-1990s, when the residential market was targeted. There was a proliferation of tariff packages, and this flurry of new offerings came about through new entry and as new capacity became available. Since operators could not perfectly discriminate among consumers, they were very careful in designing new tariffs, taking into account the effects on existing subscribers.[49]

Starting from the late 1990s, mobile telecommunications services have been supplied to the mass market and can be now considered as a commodity. The trend toward flexible pricing packages is continuing; the most

[47] See, for example, the experience in the UK as described in Geroski, Thomson and Tooker (1989) and Valletti and Cave (1998).
[48] Parker and Röller (1997) find that prices were considerably above competitive duopoly levels in US markets (prices included a 35 per cent mark-up over marginal costs).
[49] On the problem of designing contracts that can screen consumers with multidimensional preferences in a competitive environment, see Armstrong and Vickers (1999) and Rochet and Stole (2000).

important tariff innovation in this stage has been *pre-paid schemes*, which sell blocks of airtime in advance of use.[50] Pre-paid services have become the staple of the subscriber base of most mobile operators, especially in Europe. The attractiveness of pre-paid schemes stems from several factors: they give the user the ability to control expenditures, customer acquisition and billing costs are smaller for mobile operators, and there is less scope for fraud and bad debts.

While the majority of countries have adopted a system whereby the person initiating the call bears the entire cost of the call (this is referred to as 'the calling party pays', or CPP), there are a few important exceptions (Canada and the USA) where the receiver also directly contributes to the cost of each call (usually referred to as 'the receiving party pays', or RPP). Under a RPP system, consumers might be more reluctant to subscribe to mobile services because they have less control on expenses. Usage patterns may be affected, too, as users, in order to avoid paying for unwanted incoming calls, may keep their handsets switched off, or may be less inclined to give away their number. These considerations help to explain why digital mobile services developed faster in Europe than in the USA.[51] RPP pricing also makes pre-paid cards less favourable to some consumers, since it eliminates most of the appeal of such schemes, (i.e. budget control is reduced).[52] Both Canada and the USA have reviewed RPP, removing some regulatory barriers to the introduction of CPP (mainly notification procedures for users and billing systems). However, it is not clear if CPP will be introduced commercially as an optional pricing structure, and initial trials have not proved very successful.[53] Operators in the USA have already responded by launching 'bucket plans', where the customer buys monthly 'buckets' of minutes on a nationwide network and typically pays a single rate wherever the call is placed and regardless of where the call is terminated. These alternative plans may reduce the need to adopt tariffs based on CPP.

---

[50] The first pre-paid cards were introduced in Germany and Switzerland in 1995, but they were not rechargeable. The first commercial success was arguably due to the marketing strategies of Telecom Italia Mobile (TIM) that adopted rechargeable cards in 1996. By June 1999, 80 per cent of TIM's total users and almost all new users were pre-paid. On the other hand, by then only 6 per cent of users were subscribing to pre-paid plans in the USA (FCC, 2000). This discrepancy is mainly to the access pricing regime, as will be seen in chapter 5.
[51] Several international organisations such as ITU (1999) and OECD (2000) have pointed out this possibility.
[52] However, a RPP system has the advantage of *cost transparency*, as the calling party exactly knows how much it has to pay, a feature very often blurred in the CPP regime. This puts more pressure on operators to cut charges for call termination, since both incoming and outgoing calls are paid by the person that chooses the mobile network operator. This issue will be developed in more detail in chapter 5.
[53] Interestingly, in Mexico a regulatory decision introduced CPP from May 1999, the country having adopted RPP until then. The introduction of CPP coincided with record growth.

2.6 Trends in cost

As is typical for network industries, mobile telecommunications is characterised by substantial economies of scale[54] and scope. The exploitation of such economies is crucial for the spread of the services. Mobile telecommunications services became increasingly affordable because of favourable developments in the cost of equipment and service provision. The principle of Moore's law – which is a basic proposition in the semiconductor industry,[55] claiming that the performance of products double every eighteen months – also applies to the mobile telecommunications industry, as a large part of the infrastructure is based on electronic equipment. The adoption barrier represented by equipment cost has thus declined dramatically over time.

The determination of the cost of service provision has become an important issue in the industry as firms are for regulatory reasons required to align their interconnection tariffs to their underlying costs. The pervasiveness of joint costs makes cost allocation difficult. This can be illustrated by the following example. Suppose that a mobile telecommunications services firm delivers only two services: call origination and call termination. Assume also that the traffic volume for termination and origination is the same. To provide either of them it has to incur the joint cost of rolling out a network, a (cost for coverage), indicated in figure 2.10. The incremental cost of traffic termination and origination are, respectively, b and c. The stand-alone cost for termination would thus be $a+b$. Likewise, the stand-alone cost for origination would be $a+c$. The stand-alone cost for the whole network – i.e. coverage and traffic – is $a+b+c$, which is also the incremental cost for the whole network. But just because the incremental cost of the whole network is $a+b+c$, even with balanced traffic it is not correct to claim that the incremental cost of termination is $(a+b+c)/2$. The incremental cost of termination is still b. From this, it becomes immediately clear that it is of utmost importance how the cost model is set up, and what are the items registered under joint costs (see also chapter 5).

However, not all cost elements relevant in the provision of mobile telecommunications services have been declining. For instance, the more competitive environment for entry into the mobile telecommunications market has increased the licence fees for the necessary spectrum and

---

[54] The minimum efficient scale is typically at a lower level than market size, and hence more than one firm can coexist in the market. Empirical studies seem to confirm this. Foreman and Beauvais (1999) find evidence for economies of scale, though McKenzie and Small's (1998) findings are not in line with this.

[55] For an analysis of the role of the law driving diffusion of semiconductors, see Gruber (1994).

Joint costs	Coverage cost **a**	
Incremental cost	Termination cost **b**	Origination cost **c**

Figure 2.10 *Cost allocation in mobile telecommunications*

increased competition in the market has required higher expenses on sales promotion. We now discuss briefly the most important cost elements.

2.6.1 Network operation cost

The main operational costs items for mobile telecommunications are network interconnection costs, maintenance costs, personnel costs and commercial costs. One can distinguish two types of operational costs: costs that the firm can largely control, and costs determined by regulatory authorities and through bargaining. With respect the first type, mobile telecommunications firms are relatively well placed compared to fixed line firms. Increasing automation and centralisation of network management and customer care functions (e.g. automatic call distribution, interactive voice response systems) allow firms to become more efficient and reduce the associated operating costs. Mobile telecommunications firms have turned out to be much more efficient than fixed line operators with this respect. A rough measure of labour productivity in the sector is the number of subscriber lines per employee: this was 310 for mobile telecommunications and 205 for fixed line telecommunications (average for OECD countries, 1999).[56] There is also significant variance in the efficiency of the mobile telecommunications firms across countries. The countries with the highest number of mobile subscribers per employee are Spain (610 subscribers/employee) and Italy (580 subscribers/employee), whereas at the lower end are Ireland and Poland with, respectively, 220 and 180 subscribers/employee.

Concerning costs determined by regulatory decisions, in the early days of the mobile telecommunications industry firms were frequently at a disadvantage compared to the incumbent fixed telecommunications operator. A typical example is the provision of backbone transmission infrastructure. In many countries, the interests of the incumbent telecommunications monopoly heavily influenced sector-specific regulation because the government mostly owned the incumbent. Only in the 1990s did this change as sector liberalisation necessitated the appointment of an independent regulator. This meant that there was also huge scope for cost reduction by removing

---

[56] For details, see OECD (2001).

the favourable treatment for the incumbent and allowing mobile phone firms to provide their own trunk infrastructure to carry calls from base stations to control stations and switching centres. As soon as mobile firms could build their own long distance infrastructure or lease lines from third-party infrastructure providers, such as railroad and other public utility firms, a significant downward pressure on prices for leased lines charged by the incumbent fixed line firms set in. Nevertheless, a huge difference persisted among countries on the cost of leased lines. Figure 2.11 illustrates the cost of leased lines across EU countries. Portugal, the most expensive country, charged €13,450 in 1997, whereas Finland, the least expensive, charged only €2500. Competition in the supply of network infrastructure by alternative network operators dramatically reduced the cost of leased lines in many countries: leased line cost was the lowest in the most liberalised countries such as Finland, which also had the highest penetration rate for mobile telecommunications. Portugal, the country with the highest leased line costs, had tariffs up to five times higher than Finland.

Interconnection is a major cost (and revenue) item for mobile telecommunications. The well-documented comparison across EU countries may give an indication of how much such costs may vary in practice. Table 2.6 lists the interconnection cost at local level for the EU member countries and compares them with what the Commission defines as the 'current best practice', or benchmark rates.[57] Huge variations could be observed across countries in 1997, with Ireland and Austria showing multiples of the benchmark rates. Denmark, France, Germany, the Netherlands and the UK were at or even below the benchmark level of 1 Euro cent. High local interconnection rates prevailed in countries with high local interconnection rates and these effects are exacerbated with mobile interconnection tariffs. High interconnection cost countries such as Austria and Italy have even higher interconnection rates for mobile services.

2.6.2 Handset subsidies and other subscriber acquisition costs

Subsidising handsets to new subscribers is essentially a means for lowering the subscriber's entry costs to the mobile phone market: in several countries, this has been very important in pushing penetration rates. However, handset subsidies are not necessary to achieve high penetration rates. For instance, Finland had Europe's highest penetration rate for a long time, but has never experienced any large-scale handset subsidy. Operators often

---

[57] The EU Commission established 'benchmark rates' in order to force compliance with the prevailing EU regulation requiring cost oriented interconnection pricing. The track record of these efforts is well documented in the annual *Implementation Reports* published by the European Commission.

Table 2.6 *Interconnection tariffs and deviations from best practice*

	Local fixed to fixed tariff[a]	Deviation from 'current best practice'	Mobile to fixed tariff[a]	Deviation from 'current best practice'
Austria	3.3	226	8.1	710
Belgium	1.1	14	1.1	14
Denmark	1.0	−2	1.0	0
Finland	1.8	1	1.8	81
France	0.7	−30	0.7	−30
Germany	1.0	0	n.a.	n.a.
Ireland	7.0	600	7.0	600
Italy	1.5	54	4.1	312
Netherlands	1.0	0	n.a.	n.a.
Portugal	1.3	25	n.a.	n.a.
Spain	1.5	51	n.a.	n.a.
Sweden	1.1	14	1.1	14
UK	0.6	−36	0.6	−34

Notes: [a] Tariffs are in Euro cents (1997), deviations are in per cent.
n.a. = Not available
Source: EU Commission.

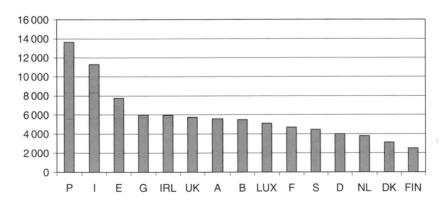

Figure 2.11 *Leased line tariffs, Europe, 1997*
The prices are in Euro and refer to the monthly price of lines of 250 km length with a transmission capacity of 2 Mbit/s
Source: EU Commission.

see this subsidy as an unnecessary cost, that has to be incurred only because the competitor is doing it, and the subscriber acquisition cost becomes very high as a result. On balance, handset subsidies seem to be a negative element for developing the market and operators who give subsidies find themselves in a *prisoner's dilemma*: they do it simply because the competitor does it, but it is actually in no-one's interest to do it. However, as long as mobile telecommunications firms receive a large portion of their revenues from high termination tariffs they have an incentive to recruit new subscribers and hence to subsidise their acquisition.

Other costs are related to the turnover of subscribers, or the 'churn'. The 'churn' rate is typically 20–25 per cent of existing subscribers within a year, which implies that a subscriber will leave the network on average after four–five years. However, 'churn' rates vary a great deal across both countries and operators. In general, the cost of gaining a new subscriber is becoming so high that operators spend considerable resources on keeping existing subscribers loyal. Subscriber 'churn' also depends very much on handset subsidies as they may attract people unable to pay the bill or provide an incentive to switch operators frequently. In Italy and Finland, where handsets are not subsidised, the 'churn' rate was around 10–15 per cent of subscribers during the second half of the 1990s. In the UK, where handsets are heavily subsidised, 'churn' rates approach 30 per cent.[58]

2.6.3 Investment cost

Radio transmission and switching is the key equipment for a mobile telecommunications network. The contributions of the various types of network elements vary with geography, the distribution of the population and the objectives of the operator. The main investment lies with the equipment related to the radio transmission between handset and network: in general, base stations account for more than 50 per cent of the cost of a network. Early industry estimates suggested that total investment per subscriber for an analogue mobile system was about $900. On top of this, a user would need to invest about $1500 for the handset.[59] Total cost per cellular mobile telecommunications user was thus initially about $2,400. However, the price declined very rapidly, and by 1985 the total cost was about half of this.[60] With the switch to digital technology, in particular to GSM operating in the 900 MHz frequency range, the investment cost per user in infrastructure declined even further, as digital technology could accommodate more users and thus better exploit economies of scale.

[58] These 'churn' rates are reported in Salomon Brothers (1997).
[59] See Blackstone and Ware (1978).
[60] See Economic Commission for Europe (1987).

Towards the end of 1990, the cost of a typical GSM network operating in the 1800 MHz frequency range was about 1.8 times the cost of a GSM 900 network with equivalent features.[61] However, the cost disadvantage of GSM 1800 diminished very rapidly with the decline in the cost of equipment. The main cost disadvantages for GSM 1800 derive from the fact that, due to the higher frequency, base stations have to be built at closer distances (i.e. not in excess of 10 km). On the other hand, GSM 1800 has the greater advantage of capacity as three times more frequencies have been allocated to this frequency. By 2000, infrastructure investment cost per subscriber had stabilised around €350, although there was considerable variance across firms and countries.[62] Moreover handset costs continued to drop and were by then already available for less than €100.

2.6.4 Licence fees

While the cost of the main equipment and operating costs were declining, an opposite movement could be observed in licence fees. In the past, governments used to assign radio frequencies with nominal fees to operators. Several countries then started to charge substantial up-front licence fee payments. The economic rationale for this was in part the reimbursement for costs sustained for making the spectrum available (e.g. removing previous users) and in part as a tax on a publicly owned good. The auctioning of radio frequencies was a device to ensure that scarce resources were allocated to the most productive purpose. This issue is developed in detail in chapters 6 and 7.

2.7 Regulation

2.7.1 Evolution of regulation

In the past, telecommunications networks were considered as 'natural monopolies' that should be operated by a single, fully integrated firm. Both economic theory and political practice recognise the need to watch monopolies closely, in particular when they are privately owned.[63] There are two different mechanisms for doing this: first, through sector-specific *ex ante* regulatory agencies; second, through general competition authorities that guard against abuses of market power in all sectors. In practice, government-owned monopolies (for instance, in most European countries) were subject to direct political control, while private monopolies (such as

---

[61] See H. Gruber and M. Hoenicke (1998).
[62] See Dresdner Kleinwort Wasserstein (2002).
[63] For a historical survey of such issues, see Gerardin and Kerf (2003).

in the USA or Canada) faced regulation by specialised regulatory agencies. This monopoly position of telecommunications services was the norm until the mid-1980s. Since then, starting in the USA and UK, a reform movement set in with a double purpose: first, to privatise state-owned firms and, second, to provide a regulatory framework that would open up the sector for competitive entry. A basic principle of regulatory reform has been the separation of the *regulatory* from the *operational* function. Ownership objectives and enforcement of impartial regulation can come into conflict when non-discriminatory decisions need to be taken. This drove the calls for privatisation of state-owned telecommunications firms to enhance the credibility of the sector liberalisation process. The principles of regulation should then be based on the criteria of independence, transparency and consultation.[64]

The role of regulatory safeguards in the transition from a monopoly to a competitive framework is to prevent incumbents or firms with market dominance exploiting their market power in order to gain unfair advantages in related markets where competition is already in place or is about to become competitive. In certain cases, it is considered as necessary to impose asymmetric terms of regulation, which means that certain provisions apply only to the incumbent, not to the new entrants. This bias may be necessary until effective competition has developed. Access safeguards typically cover issues such as: access to and use of leased lines; non-discriminatory access to infrastructure and basic services; interconnection; access to relevant information; and, last but not least, cost oriented pricing. The predominant regulatory orientation foresees a gradual evolution of regulatory tasks in the context of sector liberalisation. As markets become competitive, regulation should change from asymmetric to symmetric, creating a 'level playing field' by maintaining the overall characteristics of being transparent, at the minimum level necessary and as simple as practicable. This leads to the question of whether sector-specific regulation need be retained at all. As markets become more competitive, it should be possible to reduce sector-specific regulation and rely on competition law.

The USA is generally considered the pioneering country in telecommunications market liberalisation. The process began in 1956, when the principle of competitive entry was established by the courts with the *Hush-A-Phone Case* (Brock, 2002). In 1982, with the break-up of the Bell monopoly, most markets segments, and in particular the long-distance telecommunications sector, were liberalised, but local fixed telecommunications remained monopolies of regional firms. The liberalisation of local

---

[64] For a survey of the liberalisation trends, see OECD (1997) and ITU (2002b).

fixed telecommunications had to await the 1996 Telecommunications Act. For mobile telecommunications, the 1996 Act contained no innovations except the so-called 'reciprocal' imposition on fixed to mobile (FTM) interconnection. This measure allowed mobile firms to enjoy the same interconnection rates with the fixed line incumbent as a new fixed line entrant would have.

The UK, another pioneering country in sector liberalisation, followed a different model, based on converting a monopoly for fixed telecommunications into a duopoly. From the period 1981–91, the incumbent (British Telecom) and the new entrant (Mercury) enjoyed a protected duopoly for fixed line telecommunications, subject to a price cap regulation established by the independent regulatory authority, Oftel.[65] Similarly a duopoly in mobile telecommunications between Cellnet (a subsidiary of British Telecom) and Vodafone was established. The duopoly experiment did not lead to the desired results. In mobile telecommunications, prices remained high raising suspicions of collusion, [66] whereas in the market for fixed line telecommunications services Mercury did not manage to get a significant market share. Thus the UK government announced in a 1991 White Paper that it would consider ending the duopoly and allow further market entry. In 1993, two more mobile licences were issued and in 1996 the fixed telecommunications sector was completely liberalised, two years ahead of the target date for European-wide liberalisation set by the European Commission.

A similar approach of temporary duopoly was also adopted in Australia. Until 1992, telecommunications services were provided under the form of a state-owned monopoly. In 1993 the firm Optus was given fixed and mobile licences to compete with the monopolist Telstra in both fixed and mobile telecommunications services. For mobile telecommunications services, a third licence was also issued to Vodafone. But here, as in the UK, the duopoly in fixed telecommunications did not live up to the expectations of creating a competitive market outcome. With the 1997 Telecommunications Act, the sector was fully liberalised and since then the industry has been overseen by two main regulators – the Australian Communications Authority (ACA) and the Australian Competition and Consumer Commission (ACCC).[67]

The most radical model for liberalisation in industrialised countries was adopted in New Zealand. In 1989, all market entry restrictions were abolished; no sector-specific regulation was established and all controversies were referred to general competition law. New Zealand's approach

[65] For details see Armstrong, Cowan and Vickers (1994).
[66] See Valletti and Cave (1998).
[67] For details, see Gerardin and Kerf (2003).

thus began at the point at which the other regulatory frameworks were supposedly aiming in the long run. The reliance on competition law, however, was not sufficient to restrain the market power of the dominant firms. Protracted litigation took place, causing substantial legal expenses to industry participants. A 'Ministerial Inquiry into Telecommunications' concluded that a number of sector-specific rules were necessary.[68] A drastic change was thus introduced with a new Telecommunications Law in 2001, establishing a telecommunications-specific regulator, known as the Telecommunications Commissioner. It was felt that these additional costs of sector-specific regulation and its enforcement would be more than compensated by a reduced level of litigation.

Japan, too, is noted among the early liberalisers. The country proceeded to full liberalisation in telecommunications following the partial privatisation of the incumbent monopolist (NTT) in 1985.[69] According to the Telecommunications Business Law, telecommunications firms were differentiated between firms owning and operating facilities (Type I carriers) and firms providing services on leased facilities (Type II carriers). Type I carrier licences were more difficult to obtain and required a ministerial authorisation. Regulatory responsibility was not given to an independent regulator, but to the Ministry of Public Management, Home Affairs, Post and Telecommunications. However, tight price regulation for all firms in the market prevented consumers benefiting from liberalisation. Only in 1998 did the government introduce a tariff notification system that allowed firms to set prices without previous approval from the regulator.

The European Commission has driven market liberalisation in most of the Western European countries. The main arguments for liberalisation were the concern that a persistence of national monopolies would be counter to the principles of the Common Market, and that they would also put at risk the international competitiveness of the European information and communications industry. The main direction of EU telecommunications policy was set in 1987 with the publication of the Green Paper on the development of the common market for telecommunications services and equipment. The Commission proposed the introduction of more competition, combined with a higher degree of harmonisation, in order to maximise the opportunities offered by the Single EU market. On the basis of the favourable reaction from all market participants to the Green Paper, the Commission prepared an action programme supported

[68] For details, see Gerardin and Kerf (2003).
[69] For details, see OECD (1999).

by the Council and the other European Institutions. This programme included the following:

- Rapid full opening of the technical equipment market to competition
- Full mutual recognition of type approval for technical equipment
- Progressive opening of the telecommunications services market to competition
- Clear separation of regulatory and operational activities
- Establishment of open access conditions to networks and services through the Open network provision (ONP) programme
- Establishment of ETSI, in order to stimulate European standardisation (which had started in 1988)
- Full application of the Community's competition rules to the telecommunications sector.

These actions were implemented to a large extent through the adoption of a series of legislative measures, among which are:

- The *ONP Directive* of 1990 (90/987): harmonises the methods and conditions of public access to the network according to the principles of objectivity, transparency and non-discrimination.
- *Services Directive* of 1990 (90/388): provides for the gradual removal of special and exclusive rights granted by member states to telecommunications operators for supply of value added (by end-1990) and data services (by 1 January 1993).
- *Public Voice Telephony and Infrastructure Directive* of 1996 (96/19): calls on member states to liberalise the provision of public voice telephony services throughout the Union by 1 January 1998, while maintaining universal service. Additional transition periods of up to five years were granted to Spain, Ireland, Greece and Portugal to allow for the necessary adjustments, particularly of tariffs, and a possible additional period of two years could apply to countries with very small networks (e.g. Luxembourg), if justified.[70] In view of these deadlines, restrictions on the use of alternative infrastructure were to be lifted by 1996 and licensing and interconnection rules should be set down by 1997.
- *Mobile Communications Directive* of 1996 (96/2): abolishes special and exclusive rights in mobile communications, which was previously excluded from the Services Directive. It also aimed at an early liberalisation (by 1997) of infrastructures and the right for mobile operators directly to interconnect. It also explicitly requested member states not to refuse licences for the operation of alternative DCS 1800 mobile systems.

---

[70] In practice, all these periods turned out to be much shorter: all countries were liberalised by 2000.

The evolution of the regulatory framework in the EU should be seen as part of the wider process of economic integration in Europe initiated by the Treaty of Rome signed in 1957, which created the European Economic Community (EEC). The broader political framework of the Maastricht Treaty (or the Treaty on European Union, TEU), which entered into force in 1993, added an important new element to the legal basis for economic integration, by listing the telecommunications sector under the heading of 'Trans-European Networks'. The sector liberalisation drive culminated in the complete sector liberalisation by January 1998 under the so-called '1998 Regulatory Framework'. When this was implemented, the Commission began to design a new regulatory framework that would tackle the perceived deficiencies of the 1998 system. One of the main limits was the technological bias: the sector was essentially regulated according to the technology. This was not considered as efficient in an environment of rapid technological change. Technological convergence of previously distinct sectors such as telecommunications and broadcasting blurred traditional sector limits and created new regulatory challenges for a liberalised market. A new regulatory framework was designed that would cover the whole range of electronics communications. 'Electronic communications' in this case comprises more than telecommunications: it covers all kind of communication networks such as broadcasting and satellite networks. The new framework,[71] referring to the broader market of 'electronic communication services', has applied since 2003. A further substantial innovation of the new framework was that it moved the whole sector-specific regulation closer to competition rules. The concept of 'significant market power' had previously been used to designate those operators that enjoyed a share of more than 25 per cent of a particular market, which triggered automatically most of the regulatory obligations (such as transparency, price control, cost orientation of charges, accounting separation and non-discrimination obligations). This has been replaced by the concept of 'market dominance' taken from competition law. The practical implication is that the threshold market share triggering regulatory obligations increases significantly.[72] It also introduces the concept of 'joint dominance', which means that regulatory measures can be triggered if there is evidence of coordination among firms.[73] In

---

[71] The 'new regulatory package' was adopted by the European Parliament at the end of 2001 and entered into force as from April 2002. It was due to be implemented by member states by July 2003 and by the accession countries to the EU by May 2004. The new framework consolidated the existing European Community legislation into six Directives (Framework Directive, Authorisation Directive, Access Directive, Universal Service Directive, Data Protection Directive, Competition Directive) and one European Commission Decision on a regulatory framework for the radio spectrum.

[72] The understanding among experts is a 40–50 per cent market share.

[73] This measure primarily related to the oligopolistic mobile telecommunications market.

the new framework, it is the task of the national regulators to define the relevant markets, whereas in the old framework it was the Commission who defined them.[74] 'Relevant markets' that are candidates for regulation in mobile telecommunications are: access and call origination from mobile networks; voice call termination on mobile networks; and wholesale market for international 'roaming' on mobile networks. The new regulatory framework gives the national regulatory authorities considerable discretionary power to assess competition levels in relevant markets and apply specific regulation. It also encourages national authorities to cooperate if markets have a trans-national nature (e.g. international 'roaming'). The analysis should be based on, but not restricted by, a set of guidelines published by the European Commission.[75] However, the Commission may block any decision regarding designation, or non-designation, of operators characterised by market dominance. This should ensure that the framework remains sufficiently flexible to adapt to technological changes, and that regulatory decisions are in line with general EU policy orientations.

An international comparison of regulatory settings shows that in many countries the mobile telecommunications industry has been seen as a precursor of overall market liberalisation in the telecommunications sector.[76] The mobile telecommunications industry is subject to several levels of regulations. In practice one may distinguish between *pre-entry* and *post-entry* regulation. Pre-entry regulation concerns in particular issues of spectrum assignment, equipment standardisation and entry. Post-entry regulation refers typically to issues such as price setting, service provision and interconnection. Because of the particular spectrum requirements and the issues of equipment compatibility, pre-entry regulation is relatively strict for the mobile telecommunications industry. However, post-entry regulation of mobile telecommunications services has tended to be minimal, especially when compared to fixed telephony.

2.7.2 Pre-entry regulation

Regulations for mobile services concern, in particular, radio frequency management and technical standards. Because radio waves are not affected by national boundaries, national frequency policy has to be coordinated

---

[74] It used to be four markets: fixed voice telephony, leased lines, mobile telephony and national market for interconnection.
[75] Commission Working Document COM (2001) 175 on draft guidelines on market analysis and the calculation of significant market power.
[76] See also ITU (2002b) for a detailed review.

internationally. The regulation of radio frequencies is necessary to avoid dangerous interference and because frequencies are scarce.

The International Telecommunications Union (ITU) has established the bulk of international frequency management and technical standards, although there are additional national or supranational bodies involved. To ensure further harmonisation at the European level, the issues are dealt with by the European Conference of Postal and Telecommunications Administrations (CEPT), a body of policy makers and regulators. CEPT has handed over the responsibility for frequency issues to the European Radio Communications Committee (ERC), which also has a permanent body, the European Radio Communications Office (ERO). The European Telecommunications Standards Institute (ETSI) is mainly in charge of harmonisation and standardisation of the equipment to be used. These institutions propose voluntary standards and governments may or may not endorse those standards as mandatory (Bekkers and Smits, 1997). For North America, it is the Telecommunication Industry Association (TIA) that elaborates standards, which may then also be endorsed by the American National Standards Institute (ANSI).

Technical standards

It is up to the national regulator to decide whether or not to impose the adoption of a particular technical standard. While it seems unambiguously desirable to allow multiple competing technological systems in markets without network effects,[77] this is not evident in markets with network externalities: there are both advantages and disadvantages to having multiple systems rather than a single standard. The presence of (strong) network externalities typically leads to 'tipping' markets, where the winning technology takes the whole market. The economic literature does not provide an unambiguous answer to the question of whether the technology prevailing in the market place will also be the best one.[78] Advocates of government intervention argue that imposing a single standard makes it possible to realise network externalities faster, and reduces the technological uncertainty among consumers. Advocates of free markets point out that letting systems compete is the best guarantee to promote better technological systems (possibly a voluntary standard), and reduces the risk of being locked in to an inferior technology promoted by the government (a mandatory standard). A counterargument is that free markets may also lead to lock-in into inferior outcomes, thereby requiring

[77] The word 'system' and 'standard' may be considered as equivalent and are often used indiscriminately in the literature. However, for clarity of the argument, the word 'standard' will be used here only when it is mandatory; otherwise the word 'system' will be used.
[78] For an overview, see Katz and Shapiro (1994) and Liebowitz and Margolis (1999).

government intervention to cope with this network externality. The dispute can be rephrased as follows: one side believes that standards generate markets, while the other believes that markets generate standards.

In the context of mobile telecommunications, the main sources of network effects arise from the fact that mobile customers can use their handset only within the areas that support their technological system. Consumers who frequently travel abroad will also gain from having an international standard. Thus, depending on the mobility of consumers, network externalities can be local, national, or even international in scope. In addition to reducing consumer switching costs and creating 'roaming' possibilities, the presence of a single system also has the traditional advantage of exploiting economies of scale in the manufacturing of equipment. Shapiro and Varian (1999) argue that network effects in the cellular mobile industry are 'strong, but not overwhelming'. Empirical research related to the welfare effects of standard setting in mobile telecommunications is still in its infancy. As will be seen in chapter 4, a mandatory standard accelerates the diffusion of mobile telecommunications during the analogue period, lending support to the hypothesis that standardisation reduces information costs to consumers and induces more price competition by creating a 'level playing field' for firms. This link, however, has been weakened during the digital phase. Because there is no mandatory standard in the USA, the market generated the new CDMA technology which is the key building block for 3G mobile telecommunications. The cost of standard regulation is therefore the risk of locking into inferior standards.

Entry licensing

As well as radio frequency allocation and setting of technological standards, regulators have to make decisions concerning the structure of the market, in particular, by fixing the number of firms and their mode of entry. Countries have been quite varied in their decisions on these issues. With analogue technology there was a widespread scheme of national monopolies, but with digital technology a worldwide trend towards oligopoly has been observed. During the 1990s there was a tendency to increase the number of firms in the market, with many countries having four and more firms. In 2001, seventeen out of the thirty OECD countries had four or more firms in the mobile telecommunications market (OECD, 2003). Since then, however, a consolidation process has started, leading to the exit of firms in some countries or firms not building networks even though they have been awarded a licence. These issues will be taken up in chapter 7.

The cost of pre-entry regulatory failure can be quite high. The US delay in adopting cellular mobile telecommunications was largely due to regulatory and political indecision on whether to have a monopoly or duopoly structure. Hausman (1997) showed that welfare costs due to the late introduction of a new service could easily be assessed as being in the order of \$50–100 billion. Other costs may arise from failures to introduce standards: the introduction of an analogue standard in the USA (AMPS) and a digital standard in Europe (GSM) clearly helped to diffuse the technology rapidly. Local equipment producers had substantial competitive advantages as a result.[79]

2.7.3 Post-entry regulation

The arguments in favour of regulation revolve around market failure, which typically has two forms: *market power* or *externalities*. The role of post-entry regulation mainly consists in preventing the attempt of firms with market power to gain unfair advantages. This can be, for instance, achieved through price regulation for services and interconnection or other measures such as network sharing or number portability. By moving away from the monopoly framework to less concentrated market structures, scope for relaxing regulation should arise as the working of competitive market forces should set in. However, there are grounds for scepticism about the degree to which a limited number of firms can establish competitive pricing. There are typically asymmetries in the market structure, with firms that are longer in the market having a much higher market share, which may also give them strong market power. But even in the presence of a relatively symmetric market structure there is scope for *collusive behaviour*, against which post-entry regulation may constitute a countervailing power.[80]

Quite a large variation in the degree of price regulation can be observed across countries. In the USA, price regulation was carried out at the state level until the 1993 Reconciliation Act, according to which states were barred from regulating either entry or prices. The period before 1993 therefore provided an interesting laboratory for comparative studies on the effects of regulatory regimes. Shew (1994), for example, finds that prices were generally higher in states with regulated markets. Moreover, the type of regulation crucially matters: for example, rate of return regulation regimes have increased prices more than price cap regimes. Shew concludes that the threat of regulation is the most effective tool for low prices. Parker and Röller (1997) find that prices in US cellular markets are in general significantly above competitive levels and that mark-ups

---

[79] See Funk (2002) for a detailed review.
[80] In the European context, this is referred to as *ex ante* regulation. This regulates firms with market power in order to prevent abuse of market power before it has actually occurred.

significantly increase with cross-ownership and multi-market contacts.[81] This suggests that for the public interest strong enforcement of merger and anti-trust policy may be appropriate, even for geographically distinct markets defined by local licences for radio spectrum. Given that regulation has a positive effect on prices, firms may actually lobby for regulation because this facilitates collusion. Duso (2000) treats the regulatory regime as endogenous and finds that regulation in the USA would have had the right effects – i.e. in reducing prices. The problem, however, is that the 'wrong' markets have been regulated: markets with no regulation would have had lower prices than with regulation, whereas regulated markets would have had higher prices than would have been observed if they had not been regulated. The selection of regulatory regime is attributed to the lobbying activity of firms. The failure of regulation in addressing the supposed market failures due to market power is shown by the fact that regulated markets have much lower penetration rates that regulated ones. Once regulatory tasks for the cellular mobile telecommunications sector were transferred to the FCC, the focus was on deregulating as much as possible, in exchange, competition through additional entry replaced regulation as a tool for checking market power.

In the EU, regulation of the mobile telecommunications sector is the responsibility of the member state. To start with, there was no common approach, as each member state acted more or less independently. The regulatory tasks were, however, limited, as mobile telecommunications firms were typically subsidiaries of the state-owned telecommunications monopoly. Only two countries, the UK and France, had a duopoly in the analogue phase of the 1980s. But in the UK a wholesale/retail arrangement was introduced by expressly prohibiting mobile network firms from selling services directly to customers. This was the task of mobile service providers, who would purchase bulk airtime from mobile network firms at wholesale rates for resale to customers at retail prices. This provision remained in place until 1994, when two more network firms were licensed. However, little of the anticipated competition actually took place, and after a short period of price competition the two network firms settled

---

[81] Although multimarket contacts may theoretically enhance firms' abilities to tacitly collude, firms still need to develop a means to communicate and coordinate their actions. It is particularly important in practical anti-trust cases to understand *how* firms coordinate their prices. Busse (2000) shows that firms in the USA during the duopoly period used price schedules as their strategic instruments to coordinate across markets. In particular, she finds that identical price schedules set by one operator across different markets could help operators' efforts to tacitly collude. Price matching – i.e. different firms setting the same price within the same market – also increases average prices; however, price matching does not appear to be associated with multimarket contacts.

down to a regime of parallel pricing which remained stable from 1987 to 1992.[82]

With the increased efforts in creating a single European market for products and services during the first half of the 1990s, the Commission set up specific guidelines with which national legislation had to comply. Following the Green Paper on Mobile Telecommunications (European Commission, 1994), the principles of the EU's telecommunications policy were extended to the mobile telecommunications sector, focusing in particular on competition as the main mechanism for checking market power. As already mentioned, the Mobile Communications Directive of 1996 requested further entry of firms and abolished the mobile sector's exemption from the Services Directive. In the EU, all mobile telecommunications firms are obliged to interconnect their networks. Firms with significant market power (in practice, exceeding the market share of 25 per cent) have to publish their interconnection tariffs, which should be cost-based. To ensure the implementation of this principle, interconnection tariffs have to be cost-based and to ease implementation rates have been benchmarked by the Commission.

One of the major hurdles for competition is due to customer lock-in because of the telephone number. Numbers may be seen as a scarce public good that is conferred on firms; in the past, any customer changing telecommunications service supplier had also to change their telephone number and this switching cost could be a strong deterrent from changing supplier. The right to keep a number even when switching supplier was thus considered as an incentive for competition. However, there were conceptual problems linked to the universal application of this principle. Number portability may take different forms; originally, it was considered for the fixed network only.[83] With local number portability, only numbers within the same region with the same local code could be ported;[84] otherwise the location-specific prefix would be no longer necessary. This problem does not apply to mobile numbers, as they are not linked to a territory but have mobile network-specific prefixes. With number portability firm-specific prefixes may be no longer relevant.[85] Number portability therefore poses information problems to users, as they cannot infer from the phone number whether the phone number is a fixed or mobile one, or which firm is operating the network of the called party, and tariff

---

[82] For a detailed discussion, see Cave and Williamson (1994) and Valletti and Cave (1998).

[83] For instance, the 1998 EU regulatory framework did not foresee such a feature for the mobile sector. However, this has become mandatory in the 2002 regulatory framework. In the USA mobile number portability became mandatory from 2003.

[84] See Gans, King and Woodbridge (2001) for a welfare analysis of local number portability, and the issue of who should bear the cost of the necessary technical adaptation of the network. They conclude that it is most efficient for each firm to bear its own cost.

[85] This does not apply to the USA, where fixed and mobile numbers are indistinguishable.

transparency may be lost as a result. The trade-off may be formulated as follows. On the one hand, number portability increases competition for customers, who have lower switching costs. On the other hand, firms may have incentives to raise prices because of the higher degree of tariff ignorance by customers. The net effect is therefore ambiguous, and number portability may not be welfare enhancing.[86] Disentangling these problems pose considerable practical regulatory challenges.

Overall it can also be said that post-entry regulation remains important for the industry, and that regulatory mistakes are costly at all levels: at the pre-entry stage they are costly because of a relative market inertia; at the post-entry stage they expose customers to the market power of firms[87] or lead to increased litigation.[88] Post-entry regulation in the operation of mobile networks has, however, tended to be minimal. Thus far, mobile telecommunications firms have not been affected by broader policy concerns such as universal service. In most cases, this has been due to the fact that mobile telecommunications were considered a value added service, falling outside the regulatory scope of basic voice service.

2.8 Conclusion

This chapter has provided some basic themes central to the mobile telecommunications industry. It started with the history of mobile telecommunications technologies, emphasising the various national idiosyncrasies in the development of the industry. It then presented some stylised facts from the industry concerning the market and technological evolutions. From this emerged three key findings: first, oligopolies generated higher penetration rates than monopolies; second, mandated technical standards generated higher penetration rates than markets with non-compatible systems; third, digital systems produced higher penetration rates than analogue systems. Thus in terms of promoting diffusion, Europeans made the mistake of developing non-compatible systems and licensing mostly monopolies in 1G systems. In the second generation, the USA fell behind by not mandating or developing a uniform standard. Higher penetration rates came, however, with rapidly declining average revenues per user, which were the joint outcome of increased price competition and the move into lower-usage market segments. On the cost side,

---

[86] For a model assessing these issues, see Buehler and Haucap (2003).

[87] A different type of distortion arises when taxes are levied on the use of mobile services. Hausman (2000) finds that, due to the relative elastic demand for mobile telephony in the USA, every dollar raised in taxes imposes an efficiency loss of ¢50.

[88] Such issues are discussed, for instance, in Kahn (2004). The author reports evidence from the USA where more extensive liberalisation has led to a higher level of regulatory complexity and more scope for litigation.

technological progress reduced technically determined unit costs, but competition for customers raised market-related cost as well as certain entry costs such as licence fees. Overall, increased competition seemed to lead to a trend toward consolidation in the industry. Regulatory tasks, which were relatively reduced in the sector when compared to fixed line telecommunications, might therefore require refocusing, shifting the emphasis from pre-entry towards post-entry regulation. This will be dealt with in more detail in the chapters that follow.

3 The evolution of national markets for cellular mobile telecommunications services

3.1 Introduction

In the cellular mobile telecommunications industry, as in many other network industries, governments intervene through various types of regulation concerning important issues such as entry licensing or setting of technical standards. The governments' decisions affect the diffusion of new technologies and have important welfare implications. Moreover, the fact that national governments have adopted different policies permits an interesting comparative analysis of the performance of such policies. There is usually little consensus among economists on the optimal policies to be followed in the context of rapid market growth and fast technological change, in the early days of the industry for instance it was unclear whether the industry was a 'natural monopoly' or whether it could make room for two or more firms. It was also disputed how entry by new firms should be organised. A further issue was whether it is in the public interest that the policy maker set a technological standard, or whether this decision should be left to the market through competition among systems. Another way of putting this question is whether *standards created markets* or *markets created standards*. The development of cellular mobile telecommunications was therefore deemed to have been crucially affected by domestic policies concerning market structure and choices of technology. The main purpose of this chapter is to explore these issues in a simple analytical framework that crystallises the main points. However, to get a practical understanding of the evolution of the industry and the driving forces of its development it will be very useful to take a closer look at individual national markets.

Section 3.2 and 3.3 present the analytical framework concerning the design of market structure and setting of technological standards, showing how these decisions have evolved over time in different countries. Section 3.4 provides a representative survey of cellular mobile telecommunications markets across the globe. The insights gathered in these case studies will be useful in interpreting the results of econometric work undertaken in chapter 4. Section 3.5 draws some brief conclusions.

3.2 The analytical framework

There is an extensive literature describing the evolution of national markets in the cellular mobile telecommunications industry.[1] On the issues of pre-entry regulation, the setting of regulatory standards and entry licensing are among the most frequently quoted topics. The theoretical literature on technological standards versus competing systems in industries with network effects has grown very large and some convergence in the conclusions seems to have emerged.[2] Standardisation seems to be a socially desirable outcome:[3] for instance, standards or compatible systems tend to benefit consumers since they reduce search and switching costs. With standards, the market should therefore grow faster. This raises the question whether it is better that governments set the standard or whether it is left to market forces. Any standard, however, leads to some sort of technological lock-in,[4] which means that it is economically too costly to adopt another system. There is thus the risk that a standard, by whoever selected, will not be the most efficient one, and that it will becomes difficult to switch or develop a better one. Governments may be quite effective in establishing a standard, but this may not be the most efficient one because, for instance, of too early decisions[5] having been taken. Competition in networks very often leads to a single technology cornering the market, and the long-term coexistence of competing incompatible systems is unlikely. The economic literature abounds with examples from the computer and electronics industry.[6] Advocates of government intervention argue that imposing a standard makes it possible to reap the benefits of network externalities faster and reduces technological uncertainty among consumers. Advocates of free markets point out that system competition is the best guarantee to promote technological progress and to develop even-better technological systems. It also reduces the risk of being locked in to an inferior technology promoted by the government. A counterargument

[1] For surveys, see Hausman (2002) and Gruber and Valletti (2003).
[2] For surveys of the issues, see Grindley (1995) and Shapiro and Varian (1999).
[3] Gandal (1994) and Brynjolfsson and Kemerer (1996) provide empirical evidence for this in the context of computer software.
[4] A clear description of the problem may be found Arthur (1989).
[5] A classical case is high-definition television (HDTV). The European governments set an early European standard based on analogue technology, at a stage when the likelihood of a technological shift to a more efficient digital technology was already apparent (Farrell and Shapiro, 1992).
[6] See for instance Grindley (1995) and Gandal (2002) for references to empirical studies.

is that free markets may also lead to lock-in in inferior outcomes, thereby necessitating government intervention to cope with the network externality.[7] Definite answers on the welfare effects of market outcomes depend crucially on the market and technology parameters involved. Little work has been done in contrasting the effect of imposing standards on the diffusion of a new technology with allowing multiple systems to compete.[8]

All this begs the question as to what extent these issues concerning standardisation are relevant for the cellular mobile telecommunications industry. Once a mobile network is interconnected and has reasonably good coverage, the customers' concern about compatibility may be reduced. There are, however, benefits from standardisation that mainly originate in the provision of equipment. A widely shared standard allows for a better exploitation of economies of scale in research and development (R&D) and production of equipment. Standards with a large customer base thus tend to have lower unit costs for equipment and higher quality equipment. As seen in chapter 2, various systems were developed for both the analogue and digital cellular technology; they differ mainly in their ability to use the spectrum efficiently. The larger number of analogue systems in the early days of the cellular industry may be explained by the fact that most countries viewed cellular telecommunications as just an additional new business of the state-owned telecommunications monopoly. The development of the cellular network was thus a means of honing the innovative capabilities of national equipment suppliers. As will be shown shortly, the most successful systems (in terms of the number of adopting countries) were not necessarily the best in terms of spectrum efficiency, but rather those where the domestic market was sufficiently large (e.g. analogue AMPS for the USA) or where there was a common standard across countries (analogue NMT for Scandinavian countries or digital GSM for Europe).[9]

Concerning the design of market structure, several governments were aware from the beginning that the 'natural monopoly' argument did not

---

[7] The typical example reported in the literature is the QWERTY keyboard winning over the allegedly superior Dvorak keyboard. For a critique of the empirical relevance of network externalities, see Liebowitz and Margolis (1999).

[8] For an analysis of the presence of network effects, see Saloner and Shepard (1995). They do not directly compare competing systems with single standards.

[9] Shapiro and Varian (1999), as we have seen, argue that network externalities in the cellular mobile industry are 'strong, but not overwhelming'. For example, even if consumers are locked in to one system, they can switch to other systems at a discount in exchange for signing service contracts. They conclude that the market is not especially prone to 'tipping'– i.e. where one system takes the whole market. And, indeed, in none of the cases where competition between systems was allowed there was a system that eventually cornered the market fully and became the *de facto* standard.

Table 3.1 *The policy matrix for cellular mobile telecommunications markets*

	Single system (standard)	Multiple systems
Monopoly	1 Monopoly	2 Multiple system monopoly
Oligopoly	3 Competition in the market with same standard	4 Competition on market and technology

apply for cellular networks. Other governments reserved the mobile tele-communications industry exclusively for the incumbent fixed line telecommunications monopolist. However, in the context of worldwide sector liberalisation trends, the number of countries adhering to this principle declined rapidly during the 1990s, rising again to 78 per cent of countries in 2001 (ITC 2002a).

The 2 × 2 matrix in table 3.1 may be seen as a stylised representation of policy decisions concerning standardisation and design of market structure. This may be read as follows. In the horizontal dimension, the government decides on whether to mandate a standard[10] or to leave the issue completely to the market. In the latter case, this typically leads to competition among systems. It is an open question whether more than one system can coexist in the long run in the market, or whether the market will create a *de facto* standard.[11] For our purposes, it is assumed that coexistence of incompatible systems persists. In the vertical dimension, the policy maker decides on whether to have a monopoly or allow entry to more than one firm. Four policy combinations are thus possible. Case 1 is the typical case observed in most countries during the early phase of the cellular mobile telecommunications industry, with a monopoly and a national standard. Case 2 refers to a monopoly adopting multiple systems. This may appear a peculiar option, but nevertheless did occur in some countries at a later stage when the monopolist was provided with additional spectrum in a different frequency band. The monopolist thus took the opportunity to adopt a more efficient system, while nevertheless keeping the old system in place. As will be seen later, this happened, for instance, in Austria, where the monopolist initially selected NMT-450 and TACS later, or Italy, which

[10] This should be interpreted in the widest sense, also including the encouragement of voluntary industrial agreements.
[11] In the real world, examples of both cases may be found. For instance, in the videorecording market the VHS system took the whole market after a period of coexistence with the Betamax system. In the personal computer (PC) operating software market, the system DOS coexists with the system based on Apple Macintosh computers.

selected RTMS initially and TACS later. Case 3 indicates an oligopolistic market structure with a standard. This, for instance, is the case of the duopolies observed in the USA or UK in the analogue phase, or the oligopolies in the European countries in the digital phase, with GSM as the mandated digital standard. Case 4 may be seen as in some sense the least interventionist case, where the policy maker allows for entry and does not mandate any standard. This case, for instance, applies to the US approach with digital technology, where in the context of a competitive market structure there was no standard but competition among digital systems, leading to the coexistence of three digital systems in the market.

3.3 Empirical evidence for the policy matrix

At this stage, it may be useful to look at the empirical evidence for the decisions concerning standard and market structure. Table 3.2 represents the status of the policy matrix as of 1997. Using data provided by the ITU and other sources,[12] the table shows that of the 118 countries that had adopted an analogue cellular system by 1997, 105 had opted for a single standard, and thirteen for competing standards. A quite similar picture obtains for the countries that adopted a digital system. Of the eighty-seven countries that had adopted a digital cellular system by 1997, seventy-nine had opted for a single standard, and eight for competing standards. Several comments may be made at this point. First, while the great majority of the countries adopted standards, a fairly constant fraction of countries nevertheless adopted multiple systems, in both analogue and digital technologies. Second, while in the analogue phase monopoly was the most widespread market structure, in the digital phase oligopoly is predominant.

Let us now look at the evolution of the above findings. For instance, has there has been an evolution of the standard selection mechanism over time, passing in particular from analogue technology to digital technology? Figure 3.1 illustrates the evolution of the fraction of adopting countries that chose a single standard. The figure shows that for both analogue and digital technologies there was a single standard in all countries adopting during the first years. While countries with multiple systems of analogue technology started to appear only after eight years, this had happened after two years with digital technology. However, the share of the countries with multiple systems seems to have stabilised around 10 per cent for both technologies.

Countries have been quite varied in their decision and timing to issue first and additional licences. Table 3.2 shows that of the 118 countries that

---

[12] See Gruber and Verboven (2001b) for details.

Table 3.2 *Adoption policies for analogue/digital cellular systems, 1997*

	Single system(standard)	Multiple systems	Total
Monopoly	83/39	5/0	88/39
Oligopoly	22/40	8/8	30/48
Total	105/79	13/8	118/87

Note: The first/second number in the cells refers to the number of countries adopting analogue/digital cellular systems.
Source: Gruber and Verboven (2001b).

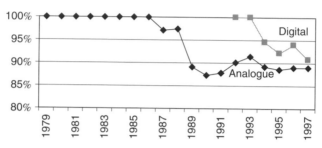

Figure 3.1 *Fraction of countries adopting a single system (standard), 1979–1997*
Note: The figures indicate the percentage of the total adopting countries that have selected a single mobile telecommunications system (standard) for analogue and digital technologies, respectively.
Source: Gruber and Verboven (2001b).

adopted an analogue cellular system, eighty-eight countries chose a mono-poly (eighty-three with a single standard and five with multiple systems) and thirty countries chose an oligopoly. This relationship was reversed for digital technology. Of the eighty-seven countries, only thirty-nine had a monopoly, whereas forty-eight had an oligopoly. This indicates a world-wide trend towards oligopoly with the introduction of digital technology. This is also confirmed in figure 3.2, which shows the evolution of the fraction of adopting countries with a monopoly for mobile telecommuni-cations. For the analogue technology, there was an eight-year period during which all the adopting countries had a monopoly; then the share of monopoly countries started to decline. For the digital technology, the pattern was the opposite. During the first two years there was no mono-poly setting at all; monopolies then set in and reached a peak of about 50 per cent after three years. The fraction of monopolies then started to decline. An attempt at an econometric explanation for these patterns will

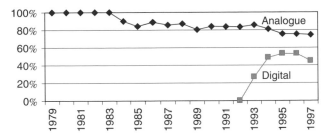

Figure 3.2 *Fraction of countries adopting a monopoly for analogue and digital systems, 1979–1997*
Note: The figures indicate the percentage of the total adopting countries that have selected a monopoly for analogue and digital mobile telecommunications services, respectively.
Source: Gruber and Verboven (2001a).

be undertaken in chapter 4; this chapter will now focus on particular elements in the main adopting countries, to in an attempt shed some light on the determinants of the evolution of the cellular mobile telecommunications industry. This will show that countries that introduce first licences early have a strong preference for a wide diffusion of the new technology. Yet during the early years of the analogue technology capacity was still heavily constrained, so that the countries gained little from introducing competition. Capacity was constrained by both technological inefficiency in terms ·of spectrum efficiency and spectrum available for cellular mobile telecommunications services. Another issue is the sequencing of market opening. When entry is simultaneous, firms obtain more or less symmetric market shares, and one would expect to compete rather 'softly'. In contrast, when entry is sequential, the entrant has to compete aggressively to obtain customers from the installed market share of the incumbent firm. A further point is the way entry licences are allocated. Countries have also chosen varied routes with different outcomes, and it may be useful to try and find some general patterns. Concerning the establishment of standards, it is also interesting to see how, in the initial phase, national standards were common, whereas with the switch to digital technology international cooperation became increasingly pervasive.

3.4 Country studies

This survey of country case studies aims to illustrate the main features that have shaped the pattern of evolution of the cellular mobile telecommunications industry. Without the intention to be comprehensive, it nevertheless attempts to show a broad range of country case studies. It focuses on aspects of policy intervention in designing market structure, and defining or even promoting technological standards. Cellular mobile telecommunications has been repeatedly declared to be a 'strategic industry' and in some countries (such as Sweden and Finland) a high proportion of manufacturing industry value added is generated in the mobile telecommunications equipment industry. Considerations of an active industrial policy have quite often driven the decisions of policy makers.[13] Typically countries that introduce first licences early have as we have seen a strong preference for a wide diffusion of the new technology. We also consider the sequencing of market opening and the way entry licences are allocated, since countries have chosen varied routes with different outcomes. Although these topics are kept in mind, the survey proceeds mainly in a chronological order for each country or region.

3.4.1 Western Europe

The telecommunications firms in several Western European countries followed the early developments of cellular mobile systems in the USA and the Nordic countries,[14] as described in chapter 2, with great interest. The firms saw the advantages of the cellular system in improving the performance of the traditional private radio systems,[15] but there were concerns about the ability of cellular systems to provide services at a low enough cost to find widespread acceptance.[16] However, Western European telecommunications companies were mainly monopolies, operators had little incentive to take risks in adopting innovations early. The initiative for developing the mobile sector was by its nature with the incumbent telecommunications monopoly operator, because mobile telecommunications were considered in the same way as fixed line telecommunications. The telecommunications monopoly was more interested in the technical feasibility of the new technology, rather than developing a business out of it. In this respect, the Nordic operators, though monopolies too, proved to be much more innovative as they had already introduced new services on the fixed network – such as ISDN, videotex and teletex – by the early 1980s. The governments of these countries were also quick to provide frequencies around 450 MHz in order to introduce cellular telecommunications services at an early stage.

---

[13] Among the most recent accounts of this, see Funk (2002) and Steinbock (2003).

[14] The Nordic countries in this context comprise Sweden, Finland, Denmark and Norway.

[15] Early technology assessments came to the conclusion that a cellular mobile telecommunications system had eleven times greater spectral efficiency compared to traditional radio systems (Frey and Lee, 1978).

[16] See Blackstone and Ware (1978) for a detailed cost comparison.

Other Western European telecommunications firms that later were to adopt cellular systems put the blame on governments in delaying the allocation of the necessary spectrum for cellular services. The World Administrative Radio Conference (WARC) decided on the allocation for mobile services in the radio frequency spectrum around 900 MHz in 1979. Soon afterwards the European Conference of Telecommunications Operators (CEPT) agreed on how this band would be used in Europe, determining matters such as the frequencies to be used for the direction of the transmission (i.e. the so-called up-link and down-link) as well as the channel spacing. It nevertheless still took some time in many countries effectively to assign this spectrum range to mobile services as it was still occupied by military applications. It was thus not until the middle of the 1980s that most of the Western European telecommunications firms were ready to introduce cellular telecommunications services. Many of these firms were late in recognising the importance of the technology because they considered cellular mobile telecommunications as a technologically very sophisticated tool, but with too small a market potential to make it worthwhile to be exploited on a large scale.

Many Western European countries introduced cellular mobile telecommunications with their own national analogue standard. In the four largest European countries – Germany, France, Italy and the UK – the telecommunications monopolies developed their own systems for mobile telecommunications and there was little interest in making them compatible; the development of the system was typically undertaken by the equipment supplier of the fixed line telecommunications monopolist. It was also the period where industrial policy was targeted at promoting 'national champions'. Only the Nordic countries realised early the advantages of a common standard. On the user side, a major benefit was expected to derive from the possibility of international 'roaming', but this was not a priority for firms at that time. From the firm's point of view, the advantage of a common standard consisted in the exploitation of economies of scale in the supply of equipment. But this was not a particular point of concern for the telecommunications monopolists in the rest of Europe. Mobile equipment production was rather seen as an attempt at honing the innovative skills of 'national champions' in the equipment industry. Even many of the countries that did not develop their own cellular system nevertheless adopted the NMT or TACS system in a very inefficient way. They introduced small variations to the system so that proprietary equipment specific to each country needed to be developed. This dissipated the economies of scale effects from adopting a widely used system. The cost to the user was increased by the fact that national markets were typically too small fully to exploit economies of scale and suffered from too little competition among mobile operators.

Table 3.3 *Introduction of analogue cellular systems, Western Europe, 1981–1990*

	System					
Year	NMT-450	NMT-900	TACS	C-450	RTMS	RC 2000
1981	Sweden Norway					
1982	Denmark Finland Spain					
1984	Austria					
1985	Netherlands Luxembourg		UK (2 systems) Ireland	Germany	Italy	France
1986		Denmark Finland Norway Sweden				
1987	Belgium	Switzerland				
1989	France (SFR)	Netherlands		Portugal		
1990			Italy Spain Austria			

Ultimately, five different analogue mobile systems were introduced in Western Europe – NMT, TACS, C-450, RTMS and RC 2000. The chronological adoption schedule is indicated in table 3.3. The most widely used system is the NMT system (in both its 450 and 900 versions), along with TACS. The C-450 system found application in two countries only, whereas RTMS and RC 2000, as we saw in chapter 2, found application in their respective home countries only. However, even if NMT, TACS and C-450 were adopted in more than one country, the systems were generally not compatible, even when they belonged to the same technology family. The Nordic countries were the exception; they jointly developed the NMT system with standardised building blocks so that equipment suppliers could compete. The system was also capable of providing 'roaming' services across the Nordic countries. Another exception was the Benelux countries, which jointly also adopted the NMT; 'roaming' within the Benelux countries was thus possible. However, the NMT system adopted was of a slightly different technical specification that made 'roaming' with the Nordic countries impossible – and, more important, precluded the economies of scale of combining equipment production with the Nordic countries. Table 3.3 also shows that until 1986 no alternative system to NMT-450 was adopted in any of these countries.

Another important feature in the development of the mobile telecommunications sector concerns market structure. In the early 1980s all countries in Europe had a state-owned monopoly for the supply of fixed line telecommunications services (for a description see Noam, 1992), and this was simply carried over to the emerging cellular sector. Exceptions with respect to the monopoly market structure during the analogue phase were the UK and Sweden. The UK government was the most consistent in introducing competition in the sector, organising a contest for the second cellular licence. The UK's duopoly approach in cellular was carried out concurrently with the country's pioneering role in liberalising the fixed line telecommunications sectors (Armstrong, Cowan and Vickers 1994). Sweden was actually the first European country to introduce competition for cellular services. However, the attribution of a second cellular licence was not well prepared – the absence of an independent regulator, for instance, led to considerable litigation (Mölleryd, 1997). Later France also joined the European group of duopolies for analogue cellular services, but the effects of the entry of a second operator had little impact on competition, since both operators were ultimately subsidiaries of state-owned companies. Price competition was also stifled because interconnection charges were kept at very high levels and hence the growth in mobile subscribers was no faster than with monopolies.

The consequences of cellular monopolies in Europe were high prices for mobile telecommunications services, reduced incentives for increasing market size and poor service performance. Figure 3.3 lists European countries in decreasing order of price for a basket of services, and at the same time shows the penetration rate for 1991, the year before digital mobile telecommunications services were introduced. One can observe a large variance across countries in the level of prices. The Nordic countries and Switzerland had the lowest tariffs. This is also partly a reflection of a policy decision to promote mobile telephony as an appropriate technology given the particular geography of the countries with their dispersed population centres. Duopolies such as in the UK and French market are not necessarily those that charge the lowest prices; low tariffs also reflect a low interconnection cost paid by the mobile firms. As most of them are subsidiaries of the fixed line telecommunications monopolist, the choice of the interconnection tariff used largely to be an arbitrary feature of internal accounting. Figure 3.3 also suggests an inverse relationship between price

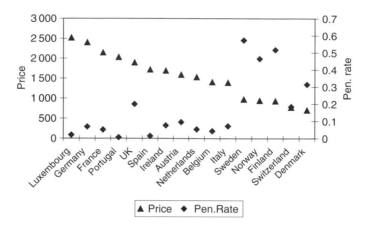

Figure 3.3 *Price of mobile telecommunications services and penetration rate, Europe, 1991*
Note: The price of mobile services is a basket in US dollars based on purchasing power parity (PPP) exchange rates (left-hand scale). The penetration rate is indicated as subscribers per 100 inhabitants (right-hand scale).
Source: OECD (1993).

and penetration rate: countries with low prices tend to have a high penetration rate and vice versa.

In spite of high prices and relatively little market growth, the first analogue cellular systems in the 450 MHz frequency range soon had problems of insufficient capacity for subscriber demand. After only a few months of operations most countries were providing additional frequency spectrum in the 900 MHz range for cellular telecommunications. Again, the exception is the UK, which did not provide spectrum in the 450 MHz range, but immediately issued spectrum in the 900 MHz range. Germany and Portugal (both using C-450), and Belgium and Luxembourg (both using NMT), on the other hand did not allocate spectrum in the 900 MHz range for analogue cellular services, therefore stifling the growth of the mobile telecommunications sector.

The allocation of frequencies in the 900 MHz frequency range required the development of a new cellular system; this was also taken as an opportunity to compare the performance of the different analogue systems in use. Given the dismal performance of purely national systems, it was no surprise that only for the relatively widely diffused NMT system was a 900 MHz version developed. National systems such as C-450, RC 2000 and RTMS were not developed further. But, more importantly, it became increasingly clear that the issue was not that of allocating more spectrum for mobile telecommunications services, but rather of utilising the spectrum more efficiently. This was achievable only by changing the technology from analogue to digital.[17]

There was the awareness in Europe that switching technology needed better coordination than in the past; this was becoming clear even before first national cellular systems were implemented, as there were already efforts under way to develop a single mobile telecommunications system for use throughout Europe. The first attempt to start a cooperative programme for a European system was made bilaterally by the French and German telecommunications monopolists in 1981. This did not go very far because of rivalry between the two different technologies, and the project was discontinued in 1982.[18] However, the idea was picked up by a wider group of European countries, laying the foundations of what would become GSM, whose establishment was ultimately the outcome of international cooperation.

The institutions behind GSM

The need for a unified European cellular mobile telecommunications standard was also recognised by CEPT (the Conférence Européenne des Postes et Télécommunications), an association of the monopoly operators for postal and telecommunications services. CEPT was a forum for the development of telecommunications standards in Europe; this is a difficult task, as there is a high risk of standard setting bodies getting locked into inefficient technologies.[19] Indeed, CEPT's past track record in the development of telecommunications standards was rather mixed. Problems typically arise from the fact that standard bodies tend to select systems on the basis of *engineering aspirations* rather than *market requirements*. One earlier failure in standardisation by CEPT was, for instance, teletex which had been intended to replace telex (Garrard, 1998). In 1982, CEPT formed a committee called Global system for mobile communications Groupe système mobile (GSM), with a mandate to specify a new mobile telecommunications system for Europe. In the light of rapidly advancing technological developments it became clear that the implications of the new system to be developed were wider: digital technology would not only enhance service performance, it would also lead the way to opening up

---

[17] See Calhoun (1988) for a detailed technical discussion on the limitations of analogue technologies and how these could be tackled by digital technology.
[18] See Garrard (1998) and Funk (2002) for a description of the issues.
[19] On the theoretical possibilities of this, and a survey of actual cases of lock-in, see Grindley (1995).

the mobile telecommunications sector to competition because of the more efficient use of the spectrum. Countries such as the UK and the USA had already shown that two firms were viable for a cellular mobile telecommunications industry, and more countries would follow this trend. The design of GSM thus happened at a time when structural change induced by liberalisation was already occurring in the European telecommunications industry. New agents such as independent regulatory authorities or competitive equipment suppliers took over roles which in most countries had been undertaken by the telecommunications monopoly. During this process, CEPT was transformed from an association of the telecommunications monopolies to a forum of national regulators. The task of establishing standards was transferred to a European Telecommunications Standards Institute (ETSI), created for this purpose. The work of GSM was formally transferred from CEPT to ETSI in 1989, maintaining the same project structure and plan.

When the working group started to develop GSM, it had the advantage that it could build an entirely new system, without taking into account of legacy systems.[20] Very few parameters were pre-set. One was that the system would operate in the 900 MHz frequency band and that it would be based on digital rather than analogue signal transmission, even though the official decision in this respect had been published only in 1987. The switch to digital technology also made it easier to find an agreement with the promoters of the different analogue technologies as the technical specifications of the digital system were less likely to privilege one of the existing analogue systems.

The requirements of the system were proposed in 1985 and widely circulated throughout the industry to provide a firm foundation for the technical details.[21] They, of course, included features such as 'roaming' and hand portable terminals. To ensure the widest possible spread of the technology, special emphasis was placed on limiting costs.

The European Commission followed the development of GSM with great interest. The overcoming of inefficient and fragmented national systems and the establishment of a pan-European standard were priorities in establishing the Common Market.[22] The European Commission

---

[20] This is different, for instance, in the USA, where the regulator required new systems to be backward-compatible with the existing AMPS standard.

[21] For a list of the requirements, see Garrard (1998).

[22] See Artis and Nixson (2001) for a description of what at that time was called the '1992 Project'. This refers to the part of the 1987 Single European Act (SEA) that committed the European Community to the completion of a single integrated market by 1992. The project brought about a revival of public interest in pushing further European economic and political integration, leading to the Maastricht Treaty, which laid the foundations for the European Monetary Union (EMU).

therefore introduced several measures to align national policies to produce the common approach that would be essential for promoting GSM. Recommendation 87/371/EEC outlined the principle of the introduction of a common digital cellular communications system throughout Europe with technical specifications similar to the system promoted by CEPT and a target introduction date of 1991. The Commission originally had a much more ambitious timetable, with the system to be implemented by 1988, which was also the original CEPT target. It became apparent at a very early stage in the programme, however, that it would be impossible to meet that target. Directive 87/372/EEC required national frequency regulators to release the necessary spectrum in the 900 MHz band. This suggests that policy intervention was useful in coordinating actions and creating the framework conditions, poor at fixing introduction times for rapidly evolving technologies.[23]

The establishment of pan-European services such as 'roaming' required much closer coordination of service modes and exchange of critical customer information. These operational and commercial issues were not covered by the technical work on the specifications, and a much broader framework for agreement on all the measures became necessary. Network operators, equipment suppliers and regulators therefore signed a Memorandum of Understanding (MoU) in 1987. The signatories committed to undertake all necessary steps to introduce GSM by 1 January 1991, a date later pushed back by six month to 1 July 1991.

Because of the complexity of the system, the introduction of GSM was ultimately delayed further, so that even the date of 1 July 1991 was not met. One of the main problems was the unavailability of user terminals; only by April 1992 were there the first type approvals for handsets. This led the way for the launch of the first network in June 1992 – the D2 network operated by a new entrant in Germany, one-and-a-half years after the original launch date.

The European Commission considered the switch to the digital technology as an opportunity to establish sector liberalisation, in line with the construction of the 1992 Single Market Project. After the establishment of GSM as a common standard in Europe, the next step concerned opening the sector to competition. The European Commission's Green Paper on Mobile Telecommunications illustrated the potential of this market for reforming the whole telecommunications sector. Directive 96/2/EC of 1996 established that the mobile telecommunications sector should be opened up by 1997, was one year ahead of the rest of the sector (for

---

[23] As will be seen, similar considerations apply to UMTS, where policy makers set an ambitious introduction date which was then not met because of technology and/or market conditions.

which liberalisation was to occur in 1998). The Directive did not specify particular entry rules for new firms. For instance, national governments were not obliged to issue mobile telecommunications licences simultaneously, but the selection criteria should be fair and non-discriminatory. All possible combinations of licensing timing (simultaneous or sequential entry) and licensing pricing (with or without a licence fee) could be followed.[24] In all EU countries the incumbent who already ran the analogue network obtained one licence; the second licence was allocated through a tender procedure (sealed bid auction or 'beauty contest'), with the price offered for the licence often being the overriding criterion of allocation. To preserve fair competition, the incumbent had to match the price paid by the winner of the contest for the second licence, or to provide some other compensatory scheme. A further provision in Directive 96/2/EC instructed member states also to issue licences for cellular mobile telecommunications services in the 1800 MHz frequency band and grant one to at least one new entrant by 1998. By that time, there were supposed to be at least three firms supplying digital cellular services in each country. Ultimately all EU countries, with the exception of Luxembourg, complied with this obligation.

Conditions of market entry

Many factors condition the entry of a new entrant into the market. Network effects suggest that 'first mover' advantages are important. The first firm in the market is usually able to attract the most profitable subscribers and therefore can quickly recover the capital costs of setting up the network. Table 3.4 illustrates for the individual EU countries the dominance of the first entrant in market share compared to that of subsequent entrants. If the second firm enters at a later stage, the first operator already has a large market share advantage. If two firms enter the market at the same time, they tend to split the market fairly evenly between them. An existing analogue customer base strengthens the position of a new entrant with digital technology, and can help to compensate for delayed entry, as existing analogue subscribers can be transferred to the incumbent's digital network. Anecdotal evidence suggests that coverage appears to be important in closing the gap between first-firm and subsequent entrants: this could be the main reason for the relatively low market share achieved by GSM 1800 firms who often, and in particular at the beginning, suffered from slow roll out, apart from the fact that they were the third entrant. Although GSM 1800 firms generally had more relaxed regulatory obligations concerning nationwide coverage,

---

[24] See Gruber and Verboven (2001a) for an econometric analysis of the determinants of the speed of diffusion.

Table 3.4 *Market shares (per cent of GSM subscribers) for mobile telecommunications firms, EU, 1997*

| | First entrant | Second entrant which enters | | Third entrant | Fourth entrant |
		At the same time as first	Later		
Austria	81		19		
Belgium	75		15		
Denmark	50	50			
Finland	35		65		
France	53		38	9	
Germany	42	45		13	
Greece	55	45			
Ireland	69		31		
Italy	69		31		
Luxembourg	100				
Netherlands	65		35		
Portugal	51	49			
Spain	64		36		
Sweden	49		34	17	
UK	32		35	15	18

Source: Author's elaboration of data from *Mobile Communications.*

they were keen to achieve such coverage as fast as possible because they perceived this as the feature that a subscriber valued above all. The most interesting case studies in the evolution of national cellular markets are now described.

3.4.2 European Nordic countries' pioneering features

The Nordic countries (Denmark, Sweden, Norway and Finland) already had a relatively advanced form of mobile communication before cellular mobile telecommunications services were introduced.[25] According to the literature, this can be attributed to a series of factors. In these countries, market growth was pushed on the supply side, as the incumbent

[25] See chapter 2. For a detailed analysis of the pre-cellular markets in the Nordic countries, see Mölleryd (1997) and Palmberg (1998).

telecommunications service firm actively promoted the use of mobile communications in each country. Moreover, governments made the necessary radio frequencies readily available for use in mobile telecommunications. The geography of most of the Nordic countries, with wide dispersion of the population in remote places, may in part explain the search for wireless solutions for access to telecommunications services, as a universal service with a fixed telecommunications network could be very expensive.

Sweden, which was the first country to introduce a mobile telecommunications service in Europe (in 1955), experienced capacity shortages by the end of the 1960s. Thus in 1969 the Nordic mobile telephone group (from which the name for the system NMT was derived) was formed, with the objective of developing a cellular mobile telecommunications system. The fact that international 'roaming' (throughout the Nordic region) was considered as important differentiated NMT from the other cellular systems that were developed elsewhere. First, this implied that the interfaces between base stations and switches had to be standardised in the whole region; this also had the advantage that it induced competition among equipment suppliers. In other countries such as the USA, only the air interface between mobile handset and base station was standardised; this absence of comprehensive standards locked mobile operators into proprietary equipment of single suppliers, with higher costs in equipment procurement.[26]

The fact that frequencies around 450 MHz rather than 900 MHz were chosen for NMT proved to be a further advantage. The 450 MHz band was ideally suited for Nordic geography, with widespread mountains and forests, since this frequency allowed economic coverage of wide areas where traffic density was unlikely to be high enough to justify the smaller cells that later 900 MHz systems would require. The disadvantage was the relatively limited capacity, due not only to the limited number of available channels but also to the greater distances required for frequency reuse.

The first NMT-450 system entered commercial service in October 1981, and the other Nordic countries started services some months later. After Japan, Sweden is credited as the second country in the world to introduce cellular services.[27]

[26] As will be seen shortly, another big difference with the USA was that in Sweden spectrum allocation for mobile telecommunications was never a problem, while in the USA operators had to lobby for a long time to receive spectrum for cellular services. The choice of 800 MHz imposed by the FCC made it more difficult and expensive to implement the system and terminals when services started. Even if authorisations had been given earlier, operators would have probably had to wait for the technology to be economically available. However, the higher frequency did allow a greater density of users without any transition from one system to another and hence proved to be an advantage in the longer term when the subscriber base was growing.

[27] Strictly speaking, Saudi Arabia introduced an NMT-450 one month before Sweden; the supplier of the network was the Swedish equipment producer Ericsson.

When NMT-450 services were launched in 1981 and 1982, coverage in all four Nordic countries was relatively limited compared with the early (pre-cellular) VHF mobile telecommunications system already in use. Significant increases in subscribers had occurred only by 1983, when coverage became fairly wide. However, on the introduction of NMT a relatively low estimate of the potential market was made. For instance, in Sweden a total of 45,000 subscribers were expected by 1991 after ten years of service, when the market was expected to reach saturation.

In fact, market growth was far beyond these forecasts. It turned out that *capacity*, and not coverage, was the key problem. Additional spectrum for cellular services was necessary to ensure that the capacity would accommodate the rising demand. Specifications were prepared during 1984 for a version of the NMT system that would operate in the 900 MHz band, as allocated by CEPT for cellular services. In the meantime, technological advances made it feasible to use these higher frequencies economically for cellular services. The NMT-900 system was very similar in terms of technical specifications to NMT-450: for instance, the two systems were intended to share the same switches. There were, however, significant improvements in other features such as reduced 'hand-over' time – i.e. the time the handsets need to jump from one channel to another during the migration from one cell to another.

All four Nordic countries simultaneously introduced the new NMT-900 service in December 1986. Hand-portable phones were introduced for the first time, greatly increasing the scope for using mobile telecommunications. The mobile telecommunications service suppliers originally wanted to provide NMT-900 coverage primarily in areas of high traffic density, where they had capacity problems with the NMT-450 system. The lower service quality due to less extensive coverage should have been offset by lower prices for NMT-900 services; it turned out that users were reluctant to take out subscriptions to the NMT-900 system because they wanted widespread coverage instead, even if they were unlikely to take advantage of it.[28] Thus the demand for NMT-450 subscriptions was increasing, forcing operators to introduce waiting lists for NMT-450, while there was slack capacity for NMT-900. Ultimately, operators had to improve the coverage of NMT-900 services; this meant they had to undertake significant additional investments in the network, which had not been foreseen originally. Because NMT-900 uses a higher frequency, the

---

[28] Coverage turned out to be paramount for mobile telecommunications customers, as several new entrants in the GSM 1800 MHz range (e.g. One2One in the UK, E-Plus in Germany and Bouygues Télécommunications in France) were very soon forced to switch from a regional to a countrywide coverage strategy to acquire customers at all, as there turned out to be a very strong preference for maximum coverage.

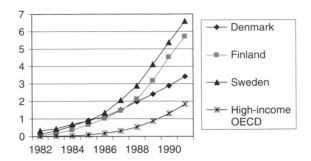

Figure 3.4 *Penetration rate for mobile telecommunications, Nordic countries, 1982–1991*
Note: The penetration rate is the number of mobile telecommunications subscribers per 100 inhabitants.
Source: ITU data.

network is composed of smaller cells and hence can provide more capacity in urban environments. On the other hand, it also needs many more base stations to cover wide, sparsely populated areas, so it becomes very costly to have a full coverage of the whole country, especially for larger countries.

Figure 3.4 shows the penetration rate, in terms of mobile telecommunications subscribers per 100 inhabitants, for the Nordic countries during the first decade of service. From this, it emerges that the Nordic countries were leading for a long time in the diffusion of mobile telecommunications, a leadership that was kept for most of the subsequent decade.

Sweden

In Sweden, the supply of telecommunications services was never a legal monopoly; in theory, anyone could set up a network to provide services in competition with dominant state-owned operator Televerket. To operate a mobile network required only an allocation of the necessary frequencies. Televerket was, however, able to create a dominant position, since it had complete control of the regulatory aspects pertaining to competition, including spectrum allocation and interconnection. That Televerket would use this prerogative to its own advantage became clear with Comvik's early experience in cellular telecommunications. Comvik was allocated only a small number of frequencies to operate a cellular network and had a market share of only 3.5 per cent in 1992. Moreover, it did not receive an NMT-900 licence. Nevertheless, because the licence to operate a 450 MHz system had been awarded to Comvik in 1981, Sweden was the first European country with a non-monopolistic mobile telecommunications market during the 1980s. The new entrant was able to start its

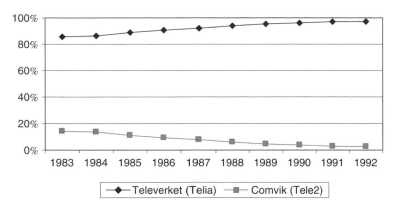

Figure 3.5 *Evolution of market shares, Swedish mobile telecommunications market, analogue phase, 1983–1992*
Note: Market share is percentage of total subscribers.
Source: Author using data from the Swedish telecommunications regulator.

service shortly before the incumbent telecommunications firm Televerket. Comvik, however, benefited only slightly from the strong growth of the Swedish market because it had adopted a proprietary technology. This decision was due to the obstructive behaviour of Televerket, which *de facto* barred Comvik from access to NMT-450 technology. Because of its proprietary technology, Comvik had more expensive terminals and never went beyond 15,000 customers, while Televerket came close to 1 million subscribers with its NMT network. Comvik had additional disadvantages deriving from the fact that it received a smaller frequency range (only twenty-eight channels out of 180). Moreover, Comvik had to adopt the receiving party pays (RPP) principle[29] for its subscribers, while Televerket did not have to. Given these asymmetries, it is not surprising that Comvik had only a very low market share of subscribers. As figure 3.5 shows, during the 1980s Comvik was unable to gain market share. It actually lost some, since Televerket gained more subscribers in a rapidly growing market. Comvik's marginal position thus persisted over time, since it did not get additional spectrum in the 900 MHz range.

Televerket initially rejected Comvik's application for frequencies to supply digital telecommunications services, too. However, an appeal to the government caused a revocation of this decision (Garrard, 1998). In

---

[29] This means that the mobile user receiving calls has to pay for incoming calls as well. As will be seen in chapter 5, this greatly affects usage behaviour. While users who do not have to pay for incoming calls can feel quite free in giving out their mobile numbers and keeping their handsets switched on, users who have to pay for incoming calls are more inhibited, for the fear of having to pay for unwanted calls.

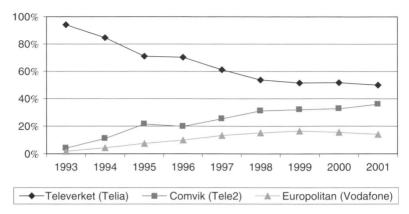

Figure 3.6 *Evolution of market shares, Swedish mobile telecommunications market, digital phase, 1993–2001*
Note: Market share is percentage of total subscribers.
Source: Author using data from the Swedish regulator.

1992, an independent regulatory body was finally entrusted with the responsibility for all telecommunications regulations, including spectrum allocation. This helped to resolve the conflict of interest of Televerket being both operator and regulator. Moreover, there was even a third GSM licence allocated to the new entrant Europolitan. So Sweden became again the leader in exploring new market structure in Europe as it was the only country in the world with three competing GSM networks in the 900 MHz range right from the start. Moreover, the dominant market position of Televerket began to be eroded only with the introduction of GSM in 1992 (see figure 3.6). It therefore appears that with an independent regulator and non-discriminatory access to radio spectrum, market shares tend to converge, at least among the two largest firms.

In 1995, four spectrum licences in the 1800 MHz range were put out to tender. This attracted four applications: three from the existing mobile telecommunications firms and one from the fixed line telecommunications firm tele8.[30] Eventually, all applicants were granted licences in the first half of 1996. The new entrant failed to live up to expectations, and was unable to construct a network. As will be seen in other cases, the Swedish market

---

[30] With twenty-three mobile subscribers per 100 inhabitants in Sweden (compared to an OECD average of seven) the penetration rate for GSM 900 networks was already the highest in the world in 1995. The issuing of a GSM 1800 licence was not primarily justified by increasing the level of competition to push the penetration rate further, as was the case in the UK and Germany, but rather to assign additional spectrum to existing operators especially for metropolitan areas where capacity limits were hit much more easily.

appeared difficult for further entry: the market showed a high penetration rate and the second and third operators were already aggressively building market share. As tele8 did not show signs of utilising the spectrum, the regulator revoked the licence in 2000. The Swedish case shows that erosion of market shares can be a very slow process, and that the existence of a dominant firm may not prejudice rapid market development. Regulatory asymmetries in favour of the dominant firms seem to have helped in sustaining a high market share: the establishment of an independent regulator seems in fact to have furthered the convergence of market shares.

Finland

Finland has a telecommunications service industry market structure that is quite unique. Though there was a monopoly on domestic long distance by Telecom Finland, for local telecommunications services there were very large number of local firms.[31] The market structure was thus highly fragmented. Most of these local telecommunications firms merged and created the firm Finnet. The government did not object to this consolidation and also helped these larger companies with mobile licensing. During the whole analogue phase of the cellular mobile telecommunications industry Telecom Finland preserved its monopoly, and was also given a GSM licence. The second GSM licence was given to Radiolinja.[32]

The extraordinary growth of mobile telecommunications subscribers was concomitant with the entry of the second operator, which made Finland a worldwide leading country in terms of mobile penetration rate. In 1995, Finland overtook Sweden in terms of penetration rate. Radiolinja anticipated Telecom Finland in introducing GSM services in January 1992, though with very limited coverage. With hindsight, this was premature and the poor service offering during the introduction phase caused losses in market share as well as reputation that the firm never recovered.

At the end of 1995, the government awarded a GSM 1800 licence to Telia,[33] the Finnish subsidiary of the Swedish firm Telia. Telia was granted considerable flexibility in its planning by the absence of any coverage target in its licence; it was therefore free to 'cherry-pick' larger cities. However, the lack of nationwide coverage became a marketing problem

---

[31] According statistics from the Finnish government, in 1938 there were 815 different telecommunications operators mainly organised as communal cooperatives. This number had declined to fifty at the beginning of the early 1990s and since then stabilised around forty-five. For a description of the Finnish telecommunications market, see Nattermann and Murphy (1998).

[32] Garrard (1998) reports that Radiolinja apparently asked the government not to put the licence out to tender, and to do so the telecommunications law had to be changed.

[33] Initially the name of the licence holder was Telivo.

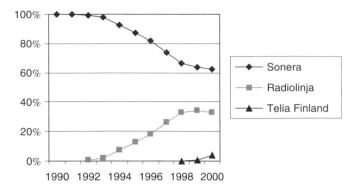

Figure 3.7 *Evolution of market shares, Finnish mobile telecommunications market, 1990–2000*
Note: Market share is percentage of total subscribers.
Source: Author, using data from the Finnish Ministry of Transport and Communications.

because users wanted such coverage, and providing full coverage to users was the only solution. Failing to reach a national 'roaming' agreement with either of the GSM firms induced Telia to sign a 'reverse roaming' agreement with the Swiss mobile telecommunications firm Swisscom.[34] This was a relatively costly arrangement, exploiting the inefficient pricing structure in mobile telecommunications,[35] and therefore could be considered as only a temporary solution. For the long term, constructing a nationwide network was necessary. The least costly way to do it would have been by the acquisition of 900 MHz frequencies, but in January 2000 Telia failed to secure the country's third GSM 900 licence, which was awarded to Suomen 2G.

Figure 3.7 shows the evolution of the market share for the three firms in the market. Until 1992 Telecom Finland (later renamed Sonera) had a monopoly.[36] During the GSM duopoly with Radiolinja, Sonera gradually lost market share, a decline halted with the entry of Telia, as this gained market share mainly at the expense of Radiolinja. Overall, the Finnish market is a case where market dominance and regulatory benevolence concerning the incumbent firm has not necessarily been a hindrance to rapid market development.

---

[34] Through this agreement, the Telia customer would be recognised as a Swiss Telecom customer, and subject to prices according to international 'roaming' agreements.
[35] For this, see also the discussion of 'inefficient bypass' in chapter 5.
[36] In 2002 Sonera, which entered at a time of financial distress following the burst of the financial market 'bubble', merged with Telia. The new firm,TeliaSonera, had to sell Telia's Finnish GSM 1800 licence; Finnet bought this, along with the network.

Along with Sweden, Finland is considered as having created a number of landmarks in the cellular mobile telecommunications sector. Beginning with a pioneering role in the development and launch of the NMT system, these include the world's first GSM service (opened by Radiolinja), and a succession of market penetration milestones. In March 1999, Finland became the first country to grant licences for 3G mobile services. Finland also became a leading supplier of equipment and the home of Nokia, the world's largest supplier of mobile telecommunications handsets.[37] The development of the mobile telecommunications market is thus considered a key instrument for industrial policy (Palmberg, 1998).

Denmark

Within the Nordic countries, Denmark was a relative laggard in the diffusion of mobile telecommunications. TDC (previously TeleDanmark), the incumbent telecommunications monopolist, introduced an NMT-450 network in 1982 and an NMT-900 network in 1987. As already seen in figure 3.4, after a take-off in line with the other Nordic countries, the Danish penetration rate evolved at a slower speed during the second half of the 1980s. A penetration rate of 3.4 per cent was achieved in 1991, the lowest mobile telecommunications penetration rate among the Nordic countries, though it was higher than in the rest of Europe.

Market growth accelerated again with the introduction of GSM, which coincided also with the entry of a second firm, Sonofon. Both TDC and Sonofon launched their networks in 1992. Sonofon began with a strategy of covering selected areas and specific customers only (such as road haulage, where the GSM service could be supplied in combination with positioning information). However, this market was too narrow to justify the large investment in a cellular network. Sonofon switched its market strategy and sought nationwide coverage to compete head-on with TDC. As figure 3.8 shows, Sonofon steadily gained market share until 1996, when it stabilised at 40 per cent. Denmark became one of the first countries pioneering the offer of combined fixed–mobile services. Both firms began to offer services based on fixed–mobile convergence in 1997;[38] the service did not prove particular popular, however, and the diffusion of combined fixed–mobile services overall remained low.

---

[37] On the history of Nokia, see Pulkkinen (1997) and Steinbock (2001).
[38] 'Fixed–mobile' convergence can have different meanings. In the Danish market, it concerned the supply of a bundle of fixed and mobile services, including features such as a single telephone number, single billing and location-sensitive pricing – i.e. the use of the mobile service was priced at the (generally cheaper) fixed line tariff when made in the 'home zone'.

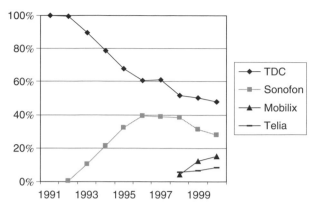

Figure 3.8 *Evolution of market shares of firms, Danish mobile telecommunications market, 1991–2000*
Note: Market share is percentage of total subscribers.
Source: Author, using data from Telestyrelsen (the Danish telecommunications regulator).

In 1997, three national licences for DCS 1800 services were awarded and one regional licence. TDC and the two new entrants, Telia Danmark and Mobilix, acquired these. The incumbent GSM firm, Sonofon, acquired only a regional licence. Mobilix, however made a national 'roaming' agreement with Sonofon to cover areas where it did not have its own network installed. Sonofon used the spectrum capacity provided by its regional licence to challenge TDC by aligning the prices for its mobile services with those for fixed line services. For Sonofon, the promotion of such convergence was a means to its goal of becoming a full-service competitor to the incumbent firm TDC. For the latter, fixed–mobile convergence was a way to cushion the inevitable loss of customers on its parent company's fixed network

The entry of two firms led to a further erosion of TDC's market share, but Sonofon experienced even greater loss in market share. Because of TDC's strong market position in fixed telecommunications services, mobile telecommunications firms claimed that that the joint fixed–mobile offering constituted unfair competition. Mobilix filed complaints with the Danish telecommunications regulator first, and later with the European Commission. Mobilix's complaint was unsuccessful, as the national regulator was unwilling to penalise TeleDanmark for introducing an innovative service. Moreover, the introduction of carrier pre-selection[39] in 1999

---

[39] This is referred to as the possibility for a fixed line telecommunications customer to access the services of selected alternative telecommunications service providers without having to dial the carrier-specific prefix code each time.

largely overrode the initial complaint, as this enables all operators to offer their own converged–fixed mobile services.

3.4.3 Spain

Spain had the policy objective of introducing cellular telecommunications services early, adopting an NMT-450 system in mid-1982, only six months after the launch of cellular services in the Nordic countries. Speed of adoption apparently overrode any considerations of cost. Cellular mobile telecommunications services worked in Spain at slightly higher frequencies than in the Nordic countries; this meant that Nordic mobile terminals could not be used for the Spanish market and therefore economies of scale in the production of mobile terminals could not be exploited. Customers could buy mobile terminals only from Telefónica, the incumbent supplier of mobile telecommunications services. This gave customers a very restricted choice. Prices of terminals and services remained at very high levels and thus the spread of mobile telecommunications was very sluggish. The service quality was also poor as territorial coverage was very limited throughout most of the 1980s. By 1990, 35 per cent of the country and 85 per cent of the population had been covered.

By that time Telefónica decided to construct a second network, switching to the TACS system. The deployment of this system was much quicker: within a year this system had provided 70 per cent population coverage, the more remarkable as TACS ran at a higher frequency and hence needed a larger number of cells. The adoption of TACS did enable Spain to benefit from economies of scale, lowering the initial entry cost by a factor of two (Garrard, 1998). Although prices for NMT-450 services were already high and contributed to sluggish market penetration, Telefónica set prices for the services from the TACS network even higher. Nevertheless subscriber growth increased with the introduction of TACS, also because of the lower terminal costs. High prices have been considered as one of the reasons of why Spain remained behind in the diffusion of cellular services in Europe.

In 1993, the Spanish government took the decision to license a second firm on the occasion of the introduction of GSM services. The licence tendering procedure, a sealed bid auction, was delayed because of domestic policy issues, but finally, the firm Airtel was awarded the licence, although it had to pay a relatively high price. Among other selection criteria for the 'beauty contest', Airtel offered to pay $670 million, which was the second highest bid; because Telefónica did not have to pay any licence fee for its GSM licence, the European Commission looked into the matter.

The European Commission's concern was directed at the possible discriminatory treatment of firms in the market, so it asked for a matching licence payment from Telefónica, or for any other type of equivalent compensating measure. During the ensuing negotiations Airtel was awarded compensation, such as a reduction of up to 50 per cent on the interconnection charges and radio spectrum to operate a GSM 1800 network in the future. When Airtel launched its first service at the end of 1995, a sharp reduction of mobile tariffs set in. Competition thus had a very important impact on prices. In 1998, the firm Retevision was awarded a GSM 1800 licence and the service was launched in 1999 under the brand name Amena; this led to a further reduction in service prices and acceleration in the penetration rate.

3.4.4 Austria

Austria was, after Spain, the second non-Nordic European country to introduce cellular services (at the end of 1984).[40] The cellular network infrastructure was based on a country-specific modification of the NMT-450 system. As in the case of Spain, special terminals were required, and hence economies of scale in the production of NMT terminals could not be exploited. The market grew steadily and by the end of the 1980s capacity problems began to arise. GSM was not ready for launch, so Mobilcom, the cellular subsidiary of the telecommunications monopolist P&T Austria, had to adopt an interim system. As in the case of Spain, this was not based on NMT-900, which would have been a sort of technological continuity; the firm justified its choice of TACS by the allegedly better suitability for hand-portable use and slightly lower cost. Domestic procurement also played a role, as the new infrastructure was commissioned from a specially created local company in which Motorola, the main supplier of TACS equipment, had a majority stake.[41]

The monopoly position of Mobilcom was never seriously questioned during the whole analogue period, not even at the beginning of the digital system.[42] As long as Austria was not a member of the EU there was no pressure to liberalise the telecommunications sector. However, once it became clear that Austria wanted to join the EU, a crash liberalisation programme had to be undertaken to adapt the telecommunications sector to the *acquis communautaire*, which implied a rapid introduction of competition into the mobile telecommunications market. This turn of events came as a surprise as by then the whole GSM frequency spectrum had already been assigned to the incumbent telecommunications operator. Accommodating a second firm required a replanning of frequency

---

[40] For a description of the Austrian mobile telecommunications market, see Pisjak (1995).
[41] Information obtained from an interview with Mr Hannes Leo from the Austrian economic research institute WIFO.
[42] See Tengg (1997).

assignment. Five consortia participated in the bidding for the second GSM licence, which was awarded to Ö-Call (later renamed Maxmobil), because it bid ATS4 billion (€290 million), until then the highest licence fee for GSM frequencies. This licence fee had to be matched by Mobilcom for the GSM licence which it had been awarded without tendering.

High licence fees tend to go with the understanding that winners of the licence have a period of exclusivity during which no other firm has similar licences. The two firms were therefore very keen on having the entry of any further firm(s) delayed as long as possible. They were also keen on insisting that any new entrant pay an amount comparable with what they had previously paid. The entry of the second operator triggered off a strong growth in terms of subscribers, which accelerated in 1997 with the entry of the first GSM 1800 firm Connect Austria, who paid ATS3.2 billion (€230 million).[43] In 1999 a further firm, tele.ring, received a GSM 1800 licence, and in 1998 Mobilcom was also given a GSM 1800 licence, in spite of the fact that the new telecommunications law (1997) should have excluded that possibility before the year 2001. Austria is a very good example of the benefits of competition and of the role of regulatory decisions in the evolution of the market.

3.4.5 The Benelux countries

The Benelux countries were the only country group in Europe outside the Nordic countries to adopt a compatible cellular mobile telecommunications system which would also allow for cross-border 'roaming'. The system was based on NMT-450, but with some differences, such as 20 kHz instead of 25 kHz channel spacing. Because of these technical changes, terminals from the Nordic regions could not be used and the Benelux countries could not benefit from the economies of scale that could otherwise have been derived by adopting the same technical specifications. Each Benelux country introduced the cellular system in a different fashion, in particular with respect to the start of service, procurement issues and the attitude towards introducing NMT-900 services. Overall, the approach was much less coordinated across the Benelux compared to the Nordic countries. Figure 3.9 shows the evolution of the penetration rate in the Benelux countries; the analogue period penetration rate has been plotted on a logarithmic scale. It shows the leading position of the Netherlands and the laggard position of Luxembourg in the analogue phase. With digital technology, Luxembourg has become the leading adopter.

The Benelux countries illustrate an interesting dilemma for standard setting bodies, especially for small countries. On one hand, they have an

---

[43] See chapter 6 for the discussion of the licence fee paid.

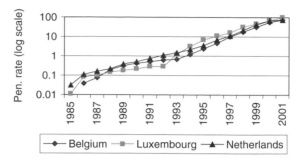

Figure 3.9 *Evolution of penetration rate, Benelux countries, 1985–2001*
Note: The penetration rate, indicated on a logarithmic scale, is calculated as number of mobile
subscribers per 100 inhabitants.
Source: ITU data.

incentive to join a common system or dominant design to benefit from
economies of scale. On the other hand, they do not want to leave all
policy variables completely predetermined by external factors. These
countries thus try to differentiate by joining a 'club' of small countries.
The question is therefore to determine the optimum size of the 'country
pool' to act as a countervailing force against dominant designs. With
hindsight, one could say that the Benelux countries would probably have
been better to join the Nordic countries, as their market turned out to be
too small to support an incompatible version of the NMT-450 system.
A description of the evolution of the market in each of the Benelux
countries is now given.

The Netherlands

The Netherlands was the first Benelux country to introduce cellular mobile telecommunications services (at the beginning of 1985). The
incumbent mobile telecommunications firm also kept a monopoly on the
supply of mobile terminals, which were available on a rental basis only.
The conditions on access to mobile telecommunications were thus highly
restrictive. The operator's monopoly over terminal supply led to a very
limited choice, since the operator bought terminals from only three manufacturers. Manufacturers had little incentive to increase the product variety given the relatively small market.

In spite of these difficulties, the number of cellular subscribers increased
steadily, and the system had reached its capacity limits by the end of 1988.
The introduction of a 900 MHz analogue system thus became necessary;
the firm learned from past mistakes and introduced a NMT-900 system in
1989, this time without country-specific modifications. This opened the

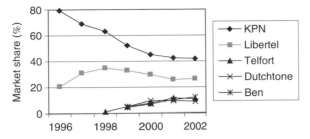

Figure 3.10 *Evolution of market shares for firms, Dutch mobile telecommunications market, 1996–2002*
Note: Market share is percentage of subscribers.
Source: Author, using data from OPTA (the Dutch telecommunications regulator).

handset market for wider choice, and subscribers were no longer subject to mandatory terminal rentals.

Although the Netherlands were quite early in the introduction of the analogue system, the switch to the digital system in mid-1994 was made after considerable regulatory delay. While the suggestion of a second GSM licence had been floated in 1991, the actual granting of the licence to the firm Libertel (later Vodafone) was not until 1995 because issues concerning the licence fee had to be resolved. The incumbent by then had a fourteen-month advantage before Libertel could start services at the end of 1995.

At the beginning of 1998, two national GSM 1800 licences were assigned to the firms Dutchtone and Telfort after an auction yielding a total of Guilders 1.8 billion (€400 million). To lower the entry barriers to the new entrants, the incumbent GSM 900 firms were obliged to offer national 'roaming' in areas where the new entrants did not have coverage. At same time, several regional GSM 1800 licences were auctioned, many being bought by the firm Ben to create a sufficiently large area to compete on a national level with the other four mobile telecommunications firms in the market. This created a unique market structure for Europe, with five firms competing in the same market. Figure 3.10 shows the evolution of the market share of the firms over time; share pattern is very interesting as it does not indicate convergence, as seen in other markets. The market shares are related to the entry into the market, with earlier entrants having larger market shares. The late entrants have similar market shares, though much lower than the early entrants.

Belgium

The Belgian introduction of analogue cellular is an illuminating example of the (negative) consequences of introducing a proprietary system into a small market and of non-competitive procurement. The deployment of the NMT-450 system in Belgium did not occur as smoothly as in the Netherlands. The main reason was that the construction of the switch was entrusted to a local company, which was unable to supply a working system before the end of 1986 (Garrard, 1998). Mobile telecommunications service could not therefore start before April 1987. The overall performance of the system was poor because the location of the base stations was driven by bureaucratic criteria than rather than by trying to identify locations best suited for network performance. Initially, there was also a monopoly in the supply of mobile terminals, but this was relaxed after six months.

The poor coverage and performance of the system in Belgium was compensated by lower tariffs than in the Netherlands; cellular subscriber growth was nevertheless slow. Unlike the Netherlands, Belgium decided not to introduce a 900 MHz analogue system, in spite of the capacity problems and the poor service to subscribers. Supply-side problems were exacerbated by the delays in the introduction of GSM. Pent-up demand for cellular services may have induced Belgacom to start GSM mobile services relatively early in 1993, actually before the Dutch. However, at the start the service quality was poor as the network functioned very badly. This reflected to a large extent the management problems with which the state-owned parent company was struggling. Relief came with partial privatisation; with the help of the US firm Airtouch, which became a shareholder, the network was redesigned to achieve satisfactory performance and services were relaunched at the beginning of 1994.

Belgium was reluctant to introduce competition in the mobile telecommunications market at a time when the GSM system was supposed to be being launched. The Belgian Parliament approved a new telecommunications law in 1991 that not only guaranteed the incumbent state-owned telecommunications firm RTT (later Belgacom) the monopoly for fixed line voice services, it extended this also to mobile telecommunications by redefining them as a basic voice service. The economic argument of the policy makers was that the country was too small to support more than one mobile firm. This line of reasoning, however, lacked foundation as several countries smaller than Belgium had demonstrated that competition was workable. There was also the successful example of duopolies in the USA from the beginning of the cellular industry, where two licences were awarded in regions much smaller than Belgium.

Belgium's approach was also in conflict with the policies of the European Commission which challenged the Belgian law, pointing out that the refusal to introduce competition in mobile telecommunications was a breach of the Treaty of Rome. The European Court of Justice (ECJ) shared this view and in October 1993 the Belgian government committed itself to introduce competition. This took some time, and only by mid-1995 was the second mobile telecommunications firm selected. The selection criterion was a competitive tendering, where cash payment carried a heavy weight; the government also set a minimum level of fee at BEF3.5 billion (€90 million). The winner out of the five competitors was the firm Mobistar, who offered to pay BEF9 billion (€225 million), which was the second highest bid. To avoid unfair competition, this price had to be matched by Belgacom for its own GSM licence. Belgacom was surprised both by the high valuation of the licence and by the fact that it had to match it.[44]

Mobistar was able to launch its services only in August 1996, a long lag. However the competitor unleashed price competition. Mobistar immediately started to undercut Belgacom's mobile tariffs substantially, to which Belgacom replied with cutting tariffs by 40 per cent. The result of this was a very rapid growth of subscribers in the Belgian market. There was also a delay in assignment of the GSM 1800 licence. The Belgian government made a third cellular telephone licence available in early 1998. KPN Orange was the sole bidder, and was awarded the licence in June 1998, with a payment of BEF7 billion (€175 million) being determined in a sealed bid auction.

Luxembourg

Initially, the incumbent telecommunications monopolist was sceptical about the economic viability of a cellular telecommunications network because of the country's small size and low-growth market expectations. Eventually an NMT-450 system was introduced, but it shared the use of a switch located in the Netherlands. Tariffs were very high, with a tariff structure different from other countries: there was no initial subscription fee but the monthly subscription was by far the highest in Europe and no distinction was made between peak and off-peak call charges. Because of overall high tariffs, market growth was so slow that capacity problems never arose, so the introduction of an interim NMT-900 system was never seriously considered.

The GSM system was introduced by a joint venture (JV) firm, with the fixed line telecommunications monopolist and the foreign firm Millicom as shareholders, in 1993. This time a national switch was built. GSM had transformed the market into a high-growth market, in spite of the monopoly. This was possible because Luxembourg was granted a temporary

---

[44] Ultimately Belgacom's payments were from the cash point of view a neutral accounting operation within the government. Belgacom had been partially privatised shortly before, with the Belgian government still holding the majority of shares. The new shareholders wanted to be compensated for this unforeseen payment by the mobile subsidiary of Belgacom. The Belgian government, as majority shareholder, also took over the payment of the minority shareholder's portion of the licence fee.

exemption in sector liberalisation by the European Commission. For many years it was thus the only country that continued to have a GSM monopoly. The announcement that the market would be opened was made in 1995, but was subject to the enactment of a new telecommunications law, which did not come into force before April 1997. At that time, a tender for a combined GSM 900/1800 licence was launched; the winner was Tango, a firm owned by Millicom International.[45] Tango started services in mid-1998. Since then, market growth has accelerated even more; by 2001, the country had reached the highest mobile telecommunications penetration rate in the world.

3.4.6 The UK

The UK was a relatively late adopter of cellular mobile telecommunications. The UK government announced its intention to grant licences for cellular mobile services in the 900 MHz frequency range only in 1982. However, the delayed start allowed the UK to learn from the experiences of cellular telecommunications in other countries. The government did not focus on technological issues but concentrated on introducing competition into the market. At that time, only Sweden and the USA had a duopoly market framework. The introduction of cellular telecommunications happened in a context when the government led by Margaret Thatcher was implementing a general privatisation and liberalisation programme for network industries in the UK, including telecommunications.[46] British Telecom (BT) was to be privatised and the fixed telecommunications sector opened up for a second operator, Mercury Communications. The newly established regulatory body Oftel regulated the sector. Given the duopoly in the fixed network, a duopoly framework for the mobile telecommunications sector was a natural consequence. British Telecom was automatically given a licence, but with a condition: mobile operations had to be run by a completely independent subsidiary in which also another partner should have a significant shareholding. BT selected the security services company Securicor as partner for the mobile telecommunications company Cellnet, which initially was called Telecom Securicor Cellular Radio (TSCR). The second licence was awarded in a 'beauty contest' in December 1982. The winner was Racal–Millicom (which later became Vodafone), a JV of the military communications equipment firm Racal and the US mobile telecommunications firm Millicom.

---

[45] Millicom was already a shareholder of the existing GSM firm. The government therefore asked Millicom to sell this shareholding.

[46] See Armstrong, Cowan and Vickers (1994) for a detailed description of the UK experience.

Formally the government did not set any standard, but it made it a condition that, whatever system was selected, it should be the same for both firms. The two firms decided on TACS, which used the US AMPS system as a base. This system had several advantages: it was proven in field trials, met the general requirements, was available from several suppliers and operated in the USA at a frequency band around 800 MHz which was not too far away from the 900 MHz used in the UK. All the other systems were less appealing. For instance NMT was at that time operating only on 450 MHz, as was C-450. The Japanese firm NTT was ruled out because there was only a single supplier.

The licence terms were identical for Cellnet and Vodafone. The government thought that the obligation to vertically separate the provision of network infrastructure from service provision to retail customers would increase competition in the market. Thus neither of the two firms was allowed to manufacture or sell equipment, or to provide value added services; services on the networks could be sold only through 'service providers'. These restrictions on vertical integration were aimed both at avoiding exploitation of a dominant market position and at providing as much competition as possible to the market at all levels: procurement, handset sales and service provision.

The service providers had a relatively simple role. They acquired customers and billed them for services, retaining a commission on the invoiced revenues. Mobile operators had to devise an incentive mechanism that balanced the conflicting interests of dealers to acquire new customers and induce customer loyalty. There were bonuses to dealers for acquiring customers, but dealers also benefited from a part of the revenues generated from these customers. Cellnet and Vodafone also competed in granting incentives to services providers, because many service providers and dealers offered the services of both mobile companies.

In spite of the ban on direct sales, either firm could set up separate subsidiaries to act as service providers, provided that these associated companies did not receive a more favourable treatment than any other independent service provider. By the end of 1985, there were thirty-nine approved service providers in the UK; out of these, twelve sold Cellnet services only, seventeen sold Vodafone services only and ten sold services from both firms. Both firms also had their own associated service provider: Racal Vodac for Vodafone and BT Mobile Phone for Cellnet. However, the service provider business was highly competitive, with a very low margin. Subsequently, a wave of consolidation set in whereby marginally profitable service providers were acquired at high prices. By 1993, the top ten service providers accounted for well over 80 per cent of the total mobile customers. Geroski, Thomson and Tooker (1989) argue that, in spite of vertical separation, mobile operators had scope for subsidising the entry of new customers (e.g. through a handset subsidy) or for raising switching costs to existing customers (for instance, through high connection charges or the requirement to change phone number when changing supplier).

Both facility-based mobile firms began services at the beginning of 1985. The deployment of the network was very rapid and the licence requirement of 90 per cent population coverage had been fulfilled a year earlier than requested. During the early phase of the market, vertical separation did not achieve the expected results in terms of enhanced subscription, and this also led to a slow diffusion of mobile services. The fact that mobile service prices stayed at the same level until 1992 was a clear indication that firms did not compete on price. Over the whole duopoly period (1985–93) both operators charged the same tariff for rental and for calls. Prices did not fall over the eight-year period and the benefits of productivity increases, which undoubtedly occurred, were not passed on to users of cellular services.[47]

To step up price competition, the UK government proposed the issuing of additional mobile telecommunications licences at the beginning of 1989.[48] The original intention was to offer mobile telecommunications services that would achieve much lower prices. With the existing cellular technology the expected cost reductions were not deemed sufficient for making mobile telecommunications a viable proposition for the much larger consumer market. The government therefore invited suggestions on how a much cheaper mobile telecommunications system could be implemented in the frequency range between 1700 and 2300 MHz. At that time, this seemed a very high frequency for any mobile service, but no more frequencies were available for mobile use in lower frequency bands. The UK government took the view that scarcity of suitable spectrum would encourage the search for new technical solutions by allocating portions of the spectrum that seemed to make their utilisation uneconomic and then wait for the technological advances to bring the cost down.

These new services were referred to as Personal communications networks (PCN). At that time the government did not expect that firms active in PCN would compete directly with the existing cellular mobile telecommunications firms, since the facilities provided by a low-cost network seemed unlikely to be sufficiently comprehensive for professional users. Cost reductions could have been achieved by omitting some important features in cellular systems such as hand-over, or by restricting coverage to urban areas or isolated cells. The driving idea behind PCN was that the

---

[47] Valletti and Cave (1998) claim that the persistence of similar pricing structures between the two firms is evidence of tacit collusion.

[48] In 1987, the Department for Trade and Industry (DTI) issued a paper titled 'Phones on the Move: Personal Communications in the 1990s', to solicit proposals.

mass market needed small, light, hand-portable phone sets, with a long battery life and low prices.

The European Commission initially criticised the unilateral action of the UK government concerning the introduction of a new system of mobile telecommunications (Garrard, 1998). The other EU countries were already cooperating to develop the new GSM system, and the UK government subsequently conceded that the cooperative option was the way to go to meet the EU criteria. So when the UK government set out the rules for tendering the two or three licences for the PCN networks, it indicated that PCN would have to be based on GSM. The existing facility-based mobile telecommunications firms and their main shareholders were excluded from the contest, since awarding PCN licences was supposed to increase the level of competition in the mobile market. The government issued three licences at the end of 1989. One went, as expected, to Mercury, the competitor of BT for fixed line telecommunications. Mercury argued that it was inhibited in effectively competing on the fixed telecommunications network with BT as long as it did not also have a mobile telecommunications licence. The regulator Oftel shared this view. The other two winners were Microtel and Unitel. The bidding occurred under considerable uncertainty, as licence winners did not know what the ultimate technical characteristics of their networks would be, as these still had to be discussed with the regulatory authorities. The UK government policy aimed at seeing the system even-tually selected widely adopted throughout Europe. Hence the standard had to be developed by ETSI, but the three UK PCN licence winners funded the preparatory works. This pioneering position gave these firms considerable influence over the direction of technical specifications. They were, however, faced with the following dilemma: the closer the standard stayed to GSM, the earlier the service would be available; but the closer it stayed to GSM the more difficult it would be to differentiate the service from conventional cellular. In the end, the changes with respect to GSM were kept to a minimum. In January 1991, the specification of what was called the Digital communications system (DCS) 1800 were announced. The only significant changes from GSM were those necessary to allow operation at a higher frequency with more channels and the limitation of handsets to lower power levels. The system would then simply be referred to as GSM 1800.

With the introduction of PCN services, the UK government dropped the obligation to use service providers for all mobile telecommunications operators, including the existing two cellular operators Cellnet and Vodafone. Moreover, to lower the entry barrier, new entrants were tem-porarily allowed to share infrastructure in rural areas (also referred to as national 'roaming'), but they should in any case cover 90 per cent of the population within ten years. Because PCN was based on the same specifications as GSM, the aim of having a cheaper mobile system than GSM 900 looked hardly achievable. The price of GSM equipment was falling steadily because of the learning curve in its production. Overall, the network cost for GSM 1800 was much larger because the system needed four–six times as many base stations as GSM for an equivalent coverage. However, GSM 1800 would then have a far greater capacity than GSM, which would deliver strong advantages in higher-density areas and the network cost per subscriber in an optimised system would eventually be lower than with GSM. Ultimately GSM 900 and GSM 1800 services had little scope for differentiation as so-called 'dual handsets' became available, and so competition turned out to be mainly on price.

The perspective of more intense competition (because PCN was technologically very close to GSM) and the high set-up cost for the network infrastructure led to the merger of Mercury and Unitel to create a firm called Mercury PCN, which later became known under the brand name One2One. The third licence holder, Microtel, changed ownership entirely, with British Aerospace selling out to the Hong Kong-based conglomerate Hutchison Whampoa, which called the firm Orange. Thus at the end there were only two PCN operators left.[49]

The entry of GSM 1800 firms definitively had a strong effect on price competition. While until 1992 the two mobile telecommunications firms kept prices at the same level, the threat of competition from PCN induced them to revise their pricing strategy (Valletti and Cave, 1998). They introduced new pricing packages especially designed for low-traffic users, by reducing the fixed charge and increasing the variable charge. This was to counter the entry of GSM 1800 firms, whose main targets were the mass market and relatively low-usage customers.

At the same time, the two existing firms were committed to deploying the GSM 900 networks which, according to the GSM Memorandum of Understanding (MoU) signed by the two cellular operators, should have occurred by January 1991. All cellular operators experienced delays on meeting this schedule. Vodafone was the most eager to fulfil the GSM commitment, and offered GSM services in July 1992, Cellnet came much later (January 1994).

If GSM 900 firms experienced delays in introducing services, GSM 1800 operators had even longer lead times. It took nearly four years from the award of the GSM 1800 licences before the first network was launched

---

[49] The cellular operators Cellnet and Vodafone received 1800 MHz spectrum in 1996. The two firms intended to use this to provide new services and to extend their reach by the use of 'dual band' handsets that would work at both 900 and 1800 MHz frequencies.

commercially. This was much longer than it had taken the holders of the GSM 900 licences that had been awarded in 1989 and the early 1990s.[50]

One2One started a service in 1993, with a business strategy to provide regional coverage only, but at lower prices. To differentiate itself from cellular operators it offered additional services such as voicemail for free. To induce high usage habits, the firm also introduced free calls during selected off peak times such as weekends.[51] Finally, to keep entry barriers as low as possible, One2One started to heavily subsidise handsets. Orange started a service in 1994, but with a strategy to provide wider geographical coverage. The greatest difference between Orange and One2One thus was the coverage strategy. Both had as a licence condition to provide 90 per cent national coverage within ten years; this was the same coverage object-ive as Cellnet and Vodafone, but with twice the time allowed to do so. The two GSM 1800 firms therefore had two different strategies: regional net-works vs. nationwide coverage. As it turned out, there was not much room for vertical product differentiation. One2One had difficulties in gaining subscribers who would be satisfied with limited coverage and eventually it had to abandon its regional coverage strategy and switch to nationwide coverage.[52]

The price cuts introduced by Cellnet and Vodafone led to a strong increase in subscribers. It also turned out that users did not notice any difference at all between GSM 1800 and GSM 900; both systems provided a mobile telephone service, and subscribers wanted reliability, high quality and comprehensive coverage at an acceptable price. Networks that tried to limit facilities or quality in return for lower cost had, at best, a limited opportunity even for basic voice services. In July 1996, any remaining boundaries between GSM 900 and GSM 1800 operators were removed when Cellnet and Vodafone were allocated radio spectrum in the 1800 MHz frequency band.

---

[50] The reason was that PCN technology took longer to develop than GSM. PCN still had several issues to resolve, such as: the development of a new system, the negotiation of licence conditions, the availability of 1800 MHz technology for base stations and handsets, the complexity of planning radio systems at frequencies that were not yet well understood; and the need to develop innovative market strategies.

[51] This strategy turned out to be very costly to the firm: it not only congested the network at certain times, it induced also large cash outflows as interconnection payments had to be made by the firm to calls directed to other firms.

[52] Valletti (1999) presents a duopoly model of firms that can decide upon the coverage of their networks. Coverage is a 'quality' parameter of differentiation. It can be shown that firms have a strong incentive to build networks of differing coverage: if the coverage requirement is sufficiently high, then equilibrium is characterised by maximal coverage differentiation. The lax initial coverage requirements may have lead to this minimum differential result for the GSM 1800 market in the UK.

Table 3.5 *Performance indicators, UK mobile telecommunications firms, 2001*

Firm	Market share of subscribers (per cent)	Market share of outgoing call revenues (per cent)	Monthly ARPU (£)	ROCE[a]
Cellnet	25	22	19	8
Vodafone	25	35	23	50
One2One	23	17	17	−23
Orange	27	26	21	10

Note: [a]ROCE: Return on capital employed.
Source: Oftel (2002).

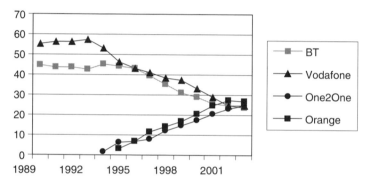

Figure 3.11 *Evolution of market shares for firms, UK mobile telecommunications market, 1989–2002*
Note: Market share is percentage of total subscribers.
Source: Author, using data from Oftel.

Figure 3.11 shows the evolution of market shares of subscribers for the firms in the UK market. The graph shows a convergence of market shares during the 1990s and by 2002 all firms had about the same market shares. The UK market is a nice example of early introduction of competition and where market shares in terms of subscribers follow predictable patterns according to simple oligopoly models. This is, however, not the case for revenues, as table 3.5 shows. In 2001 Vodafone had a market share of 35 per cent in terms of outgoing call revenues, much higher than its market share for subscribers. Inversely, One2One had a market share of 17 per cent in terms of outgoing call revenues, much below its market share for subscribers. This implies that average recurring revenue per user (ARPU) at £23 per month, is considerably higher for Vodafone than the £17 for One2One. This is also reflected in the different levels of profitability, as indicated by the return on capital employed (ROCE). Similar considerations also apply for other firms. All this suggests that there is ample scope for different strategies in terms of product differentiation. Valletti (1999) has developed a model where differentiation occurs through different degrees of network coverage. However, as will be seen in chapter 5, there are also other strategic variables that may be used for product differentiation strategies.

3.4.7 Germany

The analogue phase in the German cellular mobile telecommunications market provides an example of how technological prowess may fail to achieve success in the marketplace. The C-450 system deployed in Germany was very advanced from a technological point of view, but it turned out to be very costly and hence the market response was sluggish.[53] Siemens, the principal telecommunications equipment supplier to the telecommunications monopoly *Deutsche Bundespost* (DBP), began the development of a cellular mobile system in 1978. By 1984, both a 450 MHz and a 900 MHz system were available. DBP decided to install the C-450 system, which operated at 450 MHz; instead of a gradual deployment, DBP set a 98 per cent population coverage target right from the launch date (May 1986). DBP thus adopted a deployment strategy that was in sharp contrast with most other countries, which had used rather a gradual deployment strategy, in line with market developments.

Becoming a cellular mobile telecommunications subscriber in Germany was extremely costly. Handsets were very expensive because of the lack of competition among equipment suppliers and sophisticated and advanced technical specifications, such as SIM cards.[54] The high price of card readers contributed to making the C-450 terminals very expensive, at $7,000 during the launch phase of the service, although it had fallen to about $3,800 by 1988. Initially German mobile telecommunications service tariffs were among the highest in Europe. Compared to Nordic countries, market growth was very slow, though faster than DBP expected. Capacity

---

[53] See Garrard (1998) and Berlage and Schnöring (1995), for a description of the German market.
[54] The SIM card principle would later be used also by the GSM system. The main difference with other existing systems at that time was that the SIM card identified a *subscriber* instead of the terminal. The use of a card to contain subscriber data has a number of advantages since it provided greater flexibility and more safety features.

problems inevitably emerged, especially in large towns. The strategy of providing extensive coverage resulted in relatively large cell sizes, designed to give as large an operating area as possible with relatively few base stations. This uniform deployment strategy took too little account of traffic patterns.

One way to lower prices could have been achieved by the entry of new operators. Because DBP's monopoly was enshrined in the Constitution, introducing competition was difficult. Any attempt to reform the sector took long a time, as qualified majorities in the parliament were necessary to amend constitutional laws. Liberalisation occurred first in sectors such as satellite telecommunications services and then mobile telecommunications (in 1988) which had a marginal impact on DBP's total revenues. The entry of a competitor (to be labelled D2) was planned to coincide with the introduction of GSM; this second mobile telecommunications firm would receive a licence in competition with Deutsche Telekom, the new name of DBP's telecommunications unit, which would automatically get a GSM licence (labelled D1). In the 'beauty contest' for the second GSM licence Mannesmann Mobilfunk was determined as the winner.

Germany was the only large European country that had not introduced a 900 MHz analogue system and severe capacity problems occurred with the C-450 network. Prices remained high and the delay in the introduction of the GSM service hit German users severely. In June 1992 Mannesmann launched its GSM network covering 70 per cent of the population from the start, simultaneously with Deutsche Telekom's mobile telecommunications subsidiary T-Mobil which covered only 60 per cent of the population with its GSM network. (However T-Mobil had country-wide coverage with its C-450 system). With the advent of competition from GSM networks, the tariffs for C-450 services were sharply reduced, making them more similar to those in other countries. However, the cost disadvantage for analogue handsets persisted: GSM handsets were far cheaper from the start than those for C-450. The first GSM phone by Nokia cost DM2,700, while the cheapest C-Netz terminal sold for DM3,200. Three months later GSM handsets sold at DM1,600 (i.e. half the cost of an analogue C-450 terminal).[55] The switch to a duopoly did not have a very great effect in increasing the number of subscribers, which strengthened the belief that to have a real impact, the number of firms in the market should increase further.

The entry of a second firm in Germany almost coincided with the granting of 1800 MHz licences in the UK. As this was seen as a means for enhancing competition in the sector, such licences were the next

---

[55] This price information is from Garrard (1998).

objective of the German government, realising that would not constitute a radically new technology but was in fact a GSM system running at a higher frequency. Admitting further entrants required the government to renege on a commitment made during the previous contest for GSM 900 licences, that no further GSM licences would be granted for the following ten years. But in the end the incentive to enhance competition prevailed and in January 1992 the government announced a tender procedure for a single national licence for a GSM 1800 network, making Germany the second European country to introduce such a system.

GSM 1800 required a much larger investment than a GSM 900 network. As the new E1 system had a minimum coverage requirement of 75 per cent of the population, the deployment of the network was forecast to be much more costly than a nationwide GSM network. However, the announcement of the entry of a new firm had already produced the desired result in terms of price-cutting. In anticipation of the entry of E-Plus in 1994, the incumbent firms undertook a significant price-cutting. Nattermann (1999) studied the degree of competition in the German market, and found that all operators initially concentrated on competition in product characteristics (such as coverage) via vertical product differentiation, in order to avoid excessive price competition. E-Plus' network was initially planned according to a strategy of regional focus, but this did not work in the marketplace. As already mentioned in the UK case, this strategy of vertical differentiation could not be supported in the market, as users preferred to have access to nationwide coverage. E-Plus was thus forced to accelerate the deployment of a nationwide network. As the scope for competition in coverage levels declined (e.g. all firms achieved an equivalent level of coverage) price competition increased. Eventually the tariffs of the analogue C-Netz even fell below GSM tariffs. Despite this, the level of tariffs was generally higher than in other countries and the mobile telecommunications penetration rate, though growing rapidly, was still relatively low.

The German government thus decided to step up the level of competition in the sector. This move coincided with a period of general liberalisation of the telecommunications sector. Before issuing further licences, the government had to respect a four-year exclusivity period for the GSM 1800 licence it had granted to E-Plus. Eventually a second GSM 1800 licence was issued at the beginning of 1997 to the firm Viag Interkom. To relax price competition somewhat, the entry strategy for this relatively late entrant consisted in offering new services, such as integrated fixed mobile services. As figure 3.12 shows, with the entry of the fourth firm Viag Interkom, there was a sharp increase in subscriber penetration rate. There was also an increase in capacity as the incumbent firms with a GSM 900 network were able to bid for additional 1800 MHz spectrum in

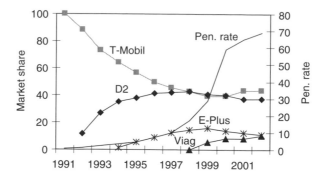

Figure 3.12 *Evolution of market shares of firms and penetration rate, German mobile telecommunications market, 1991–2001*
Note: The market share is based on the total number of subscribers. The penetration rate is calculated as the number of mobile telecommunications subscribers per 100 inhabitants.
Source: Author, using ITU and Mobile Communications data.

an auction in 1999. Looking at the evolution of market shares, one can identify a strikingly different pattern from that observed in the UK. Here, market shares do not converge to the same level for all firms, but at two different levels according to entry timing. For the two GSM 900 firms market shares converged at about 40 per cent, whereas for the later entering GSM 1800 firms market shares seem to converge at around 10 per cent. The poor profitability of these two firms, and the persistent press rumours on a possible merger between them, raises the question of the sustainability of such a market structure.

3.4.8 Italy

Italy decided to develop its own analogue cellular mobile telecommunications system, which became known as Radio telephone mobile system (RTMS). Italtel, the main local telecommunications equipment supplier, developed the system in close cooperation with SIP, the state-owned telecommunications firm. Commercial supply of cellular mobile telecommunications services had started by the end of 1985; however, with respect to several features, RTMS was not up to the technical levels of other, more widespread, systems such as NMT or AMPS. Although RTMS provided key features of cellular systems such as hand-over, the system's performance was poor.[56] RTMS also retained the old concept of calling areas used

[56] See Garrard (1998).

by the pre-cellular mobile telecommunications systems. The Italian territory was divided into ten areas and a caller therefore had to know in which
area the called mobile phone was in order to complete the call. Moreover,
hand-over was possible within, but not across, areas.

In the tradition of telecommunications monopolies, SIP had exclusivity for the supply of RTMS terminals, which were manufactured by only
three Italian manufacturers. Mobile telecommunications terminals were
therefore expensive and the range of choice limited. Nevertheless, RTMS
soon proved to be incapable to satisfy steadily growing demand. Just
45 000 subscribers were forecast within ten years of operation, whereas
100 000 subscribers had been already reached after three years (1988). An
interim system was therefore needed to satisfy demand before GSM
services could be deployed. This time, SIP did not take any chances and
decided to adopt the already proven TACS as an interim system for
cellular telecommunications in the 900 MHz band until it could be
replaced by digital systems. TACS entered service in April 1990 and
proved to be highly popular. The system was a considerable improvement over RTMS in terms of features and performance. Moreover,
Italian subscribers were able to benefit from cost reductions for terminals
because of the large number of TACS terminals already sold in other
countries, in particular the UK.

There was quite a long delay between the announcement (in 1990)
that a second GSM operator would be licensed and the actual award
of the licence (in 1994). This liberalisation step in mobile telecommunications required a complete overhaul of the fragmented Italian
telecommunications system.[57] For instance, the mobile telecommunications monopoly granted to SIP until 2003 needed to be revised. At the
same time, state-owned telecommunications activities were concentrated in one single company (Telecom Italia). In 1993, Mobile telecommunications operations were transferred to the subsidiary Telecom Italia
Mobile (TIM).

The contest for the second GSM licence was won by the consortium
Omnitel Pronto Italia, offering an up-front payment of Lire 750 billion
(€375 million) and a minimum payment of Lire 160 billion (€80 million)
for each year for five years, depending on forecast revenues. TIM had
received the licence without paying any significant licence fee; the
European Commission looked into the matter and considered the payment
requested from Omnitel as discriminatory and anti-competitive. The
European Commission requested that TIM should pay the same licence

---

[57] For descriptions of the Italian market, see Guerci *et al.* (1998) and Cambini, Ravazzi and
Valletti (2003).

fee, or that Omnitel should receive some compensation. After a lengthy dispute, the issue was settled by compensation: Omnitel received indirect compensation through a 25 per cent discount on interconnection rates charged by Telecom Italia; Omnitel was also allowed to use TIM's base stations temporarily in areas which Omnitel did not cover (national 'roaming'); and TIM was also barred from promoting the GSM network before Omnitel was ready to start services.

The delay in awarding the second GSM licence gave TIM plenty of time to fully exploit its TACS network and to introduce GSM far in advance of its new competitor. TACS growth was boosted by the introduction of special tariff packages designed for family users. In 1997 the Italian TACS network reached a peak with 3.7 million subscribers and became the single largest analogue cellular network in the world. What is surprising in the Italian case is the long time the large analogue subscriber base was supported. In many countries, the analogue network was phased out during the second half of the 1990s, in Italy analogue cellular services were still being actively marketed after the 2000, in spite of the fact that with the analogue system international 'roaming' was not possible and a series of so-called 'value added services' could not be supported either. This is even more surprising as there was no significant difference between the tariffs for analogue and digital services.

When Omnitel launched a pre-commercial trial of its GSM network in October 1995, TIM had already reached the capacity limit of the TACS network and new subscribers were being directed onto the GSM network. Omnitel declared that it did not compete with TIM on price but on service quality. As a matter of fact, the tariffs of the two firms were quite similar. However, the entry of Omnitel triggered a very rapid market growth in terms of subscriber figures. Italy overtook the UK as the largest European market in terms of subscribers at the end of 1996. The major element in the observed growth in Italy was attributed to *product innovation*, in particular new tariff formulas such as subscriptions without monthly fees and a pre-paid phone card. This card (dubbed Timmy) had been launched first by TIM and was an extraordinary success, attracting over 1 million new subscribers in the last quarter of 1996. Omnitel was taken by surprise by this move and reacted late. However, within the three months of its response (with *Libero prepagato*, in January 1997) it had attracted 250 000 new subscribers.

Pre-paid cards have advantages for both operator and user. For the operator, it reduces acquisition costs, avoids billing cost, reduces bad debt and permits the tapping of new consumer segments. On the other hand, since the average airtime charge is much higher, this card will not take away intense users. For the consumer, the main benefits are in avoiding a monthly billing tax of Lire 10 000 (€5.16) and better cost control. Overall, since the pre-paid card requires a modest one-time up-front fee of Lire 50 000 (€25.82), switching operators is not costly. Pre-paid cards do not have a monthly bill, which explains also the extraordinary growth the product had on the Italian market. At the end of 1999, around 75 per cent of the Italian cellular market base was constituted by pre-paid subscriptions, compared to 24 per cent in Germany, 50 per cent in France, 36 per cent in Sweden and less than 10 per cent in Finland. This rapid growth puzzled observers when they tried to reconcile the evidence of successful market growth with an allegedly poor industrial framework characterised by a long history of regulatory and licensing delays (PNE, 2000).

Additional competition from GSM 1800 arrived late, as the first licence of this kind to the firm Wind was awarded only at the beginning of 1998. Wind started mobile telecommunications services in March 1999, setting up both a fixed and a mobile telecommunications network infrastructure. Wind thus wanted to leverage its presence in the fixed line market to become a strong supplier in fixed/mobile convergent (FMC) technologies.

A second GSM 1800 licence was awarded after a 'beauty contest' to the firm Blu in July 1999. This firm was late in taking up services, entering the market in May 2000, when the mobile penetration rate in the Italian market already exceeded 58 subscribers per 100 inhabitants. At such a penetration rate the acquisition of new customers had to be achieved mainly by undercutting prices in order to steal customers from other firms or to attract marginal customers to the mobile telecommunications market. In both cases, adverse selection led to subscribers with low ARPU. Following Blu's failure to secure a UMTS licence in October 2000, the company was broken up in 2002 and its assets distributed amongst Italy's other market players. Blu is the first example of a failed mobile telecommunications firm in Europe and an indication that a market may not sustain more than three firms in the long run, especially with highly asymmetric market shares. Figure 3.13 shows the evolution of the market shares of the different firms and the penetration rate for Italy. It shows a large variation in market shares and advantages of early entry seem to persist. The growth in the penetration rate showed some acceleration by 1999 when Wind and then Blu entered. In 2001, the growth in penetration rate started to decline and this may make it increasingly difficult for new entrants to acquire customers, as most of the new customers will need to be taken away from existing firms. Blu's failure may therefore be due to the fact that it was entering relatively late in a market that was becoming saturated.

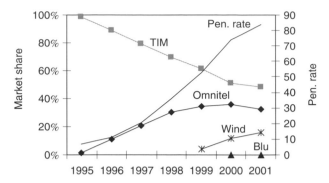

Figure 3.13 *Evolution of market shares of firms and penetration rate, Italian mobile telecommunications market, 1995–2001*
Note: The market share is based on the total number of subscribers. The penetration rate is calculated as the number of telecommunications subscribers per 100 inhabitants.
Source: Author, using ITU and Mobile Communications data.

3.4.9 France

The French telecommunications monopoly was operated, until the creation of the public company France Télécom in 1989, by the government's Direction Générale des Télécommunications (DGT). DGT took a completely different approach to the provision of mobile telecommunications services compared to firms in other countries. The firm wanted to combine the characteristics of a cellular network with that of a low-cost private mobile radio. The French military equipment supplier Matra had been charged in 1981 to develop Radiocom 2000 (RC 2000), a system that would combine these two needs. France was therefore rather late in developing a mobile system compared to other countries. Mobile telecommunications services were not introduced until 1985 when the system was launched in the Paris area only. While it was quite common for mobile telecommunications firms in Europe to introduce cellular services with such limited coverage at their initial launch, in France subsequent deployment also remained slow. By the end of 1986 there were only fifty base stations in operation, and by 1988 the network covered less than 40 per cent of French territory.[58]

Technically, the RC 2000 system was not a cellular system in the traditional sense. It was an unusual mixture of conventional radio relay techniques and some cellular principles. For instance, hand-over was built into

[58] See Garrard (1998).

the technical specifications but was not implemented for many years, and the operating frequencies were too low to allow the level of frequency reuse expected from cellular systems. The dispatch radio service could also interconnect to the fixed voice network. The operating frequencies varied substantially across areas so that terminals were not always compatible across all national regions. The services supported by the RC 2000 network were marketed with two quite different features, in line with the original technical objectives. On the one hand the system was considered as a business network in the form of a dispatch radio service with optional interconnection to the fixed networks; on the other, the system was considered as a means for providing conventional automatic mobile telecommunications services, but without hand-over facilities. The pricing structure for RC 2000 services was much more complicated than conventional cellular services, as prices varied across regions. This differential pricing was designed to discourage subscribers from using the national service and from making calls in congested areas. Mobile users had also to pay for incoming calls. This made it the only case in Europe where the receiving party pays principle (RPP) was applied.

In view of these growth-inhibiting technological factors, the negative consequences of 'national champion' industrial policies centred on protecting domestic firms became evident. Matra was initially the sole supplier of terminals. But since its production capacity was insufficient, other suppliers entered the market. However price competition was very relaxed and prices for terminals remained high. Overall, the performance of RC 2000 was poor; the system was quickly congested. Subscribers had to buy a mobile phone specifically for the coverage they wanted, but could not then change from, say, service in Paris to full national availability because the frequency used was in a completely different band and they would have to purchase a new mobile phone.

Dissatisfaction with the performance of RC 2000 induced the French government to invite tenders for the operation of a second mobile network in mid-1987. The licence was awarded to Société Française Radiotéléphone (SFR), with the state-owned water company Lyonnaise des Eaux as the main shareholder. The government had not specified any technology in the tender, so SFR selected an upgraded version of the NMT-450 system, which included some new features that had been developed for NMT-900. Despite the performance benefits of these modifications, the market-specific adaptation entailed a loss in economies of scale, and users lost the price advantages that they could have otherwise gained by remaining with an already adopted system.

The entry of a second mobile telecommunications firm did not lead to significantly enhanced price competition, because both companies were

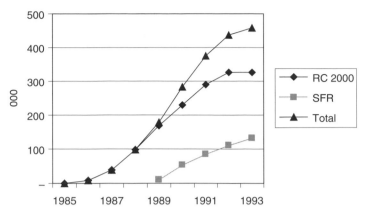

Figure 3.14 *Evolution of number of subscribers for analogue mobile telecommunications, France, 1985–1993*
Source: EMC.

ultimately state-owned. Interconnection rates were high, so SFR did not undercut RC 2000's high prices. This also meant that there was no significant acceleration in the increase of subscribers as a result of entry (see figure 3.14). Overall, the French analogue cellular market performed very poorly.[59]

It can therefore be concluded that the early attempt of stimulating market growth through market entry of a competitor was unsuccessful in France. It must be taken into account that neither firm had the necessary spectrum to provide a high-quality service to a large number of users in any case. RC 2000 had a large number of channels, but at low frequencies that could not be reused efficiently. SFR had an allocation of only 6 MHz, providing a maximum capacity of 140 000 subscribers. Because of these technical and market restrictions, SFR coverage never reached that of RC 2000, which itself was quite limited. The subscribers on the analogue network reached a peak in 1993, and declined immediately when GSM networks were launched.

During the tendering for the second analogue licence the government had indicated that the winner would also be granted a GSM licence,

---

[59] Another alleged reason for the slow market growth for cellular services was the introduction of CT-2 Telepoint services. Whereas many countries, such as Germany, Finland and the UK, had abandoned these services after an unsuccessful trial period, France stuck to them for a longer period. The service, launched under the name BiBop, was made first available in Paris, Strasbourg and Lille at its peak in 1992, reaching 100 000 subscribers by 1995. However, the decline of this technology was clearly illustrated when contrasted with the growth of GSM services.

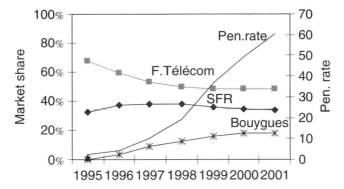

Figure 3.15 *Evolution of market shares of firms and penetration rate, mobile telecommunications market, France, 1995–2001*
Note: The market share is based on the total number of subscribers. The penetration rete is calculated as the number of mobile telecommunications subscribers per 100 inhabitants.
Source: Author, using ITU and Mobile Communications data.

subject to meeting unspecified standards of satisfactory performance. Whereas France Télécom was certain to receive a GSM licence, the definitive response to SFR was delayed. The supposed official launch dates for the GSM networks were July 1992 for both operators, but the date ultimately slipped by more than a year.

In 1992, the government started consultations on the introduction of GSM 1800 licences, deciding to issue a single GSM 1800 licence through a 'beauty contest'. Because of the relatively low population density in the country, it was clear to potential operators that apart from certain urban agglomerations, widespread coverage of the country would be very costly. Only three bids were presented. The licence was awarded to Bouygues Télécom in 1994, granting Bouygues exclusivity for GSM 1800 services for four years in five major cities. The GSM 1800 service was launched in mid-1996 in these five towns and connecting roads, covering some 15 per cent of the total French population. Low tariffs had to make up for the lack of coverage. The launch of GSM 1800 induced France Télécom to introduce for the first time a tariff package aimed at low-usage consumers. The limited coverage strategy was met with a sluggish response from users, who seemed to prefer national coverage, which was achieved in November 1998. Since then, the penetration rate has shown a significant acceleration (see figure 3.15) suggesting that the third firm provided a significant impetus for competition. The evolution of the market share seems to have stabilised, but at different levels depending on the entry date of the firm.

3.4.10 Portugal

Portugal was one of the poorest countries in Western Europe and policy gave low priority to the development of cellular telecommunications. The fact that the country received subsidies for adopting a German C-450 system certainly assisted in the decision to introduce mobile telecommunications. The system was launched late, at the beginning of 1989; the C-450 version adopted used a different channel spacing from that used in Germany. The consequence was that one of the poorest countries in Europe introduced the most expensive cellular system. On top of this, the custom-built version of C-450 made the terminals even more expensive than those in Germany at the time. Service prices were relatively high and coverage very limited; market growth was thus very slow, at least until the entry of new firms.

Concerning sector liberalisation, Portugal was a relatively early mover. A new telecommunications law was passed in 1989, leading the way for competition in mobile telecommunications.[60] In 1991, the invitation to tender for the second GSM licence was issued and the ensuing 'beauty contest' was won by Telecel, mainly on the basis that its business plan forecast 230 000 subscribers by 2000. (None of the other bidders forecast more than 90 000 subscribers.) With hindsight, all forecasts were too low as market growth was even more rapid than Telecel's estimate. Telecel started service in late 1992 and met the 2000 objective in 1994, in spite of the fact that tariffs were relatively high. The introduction of pre-paid subscriptions was the main contributor to the rapid market growth. The incumbent mobile telecommunications firm TMN launched GSM services simultaneously with Telecel, but it was less successful in attracting new subscribers, and therefore had a lower market share most of the time. In 1997, the firm Optimus was awarded a GSM 1800 licence and started service at the end of 1998. Subscriber growth turned out to be exceptionally high in Portugal and the telecommunications market accounted for a relatively large share of gross domestic product (GDP). The telecommunications sector accounted for 4.10 per cent of GDP in 1999, the highest share of GDP among OECD countries. (In comparison, the OECD average share of the telecommunications sector in GDP is 3.12 per cent, OECD, 2001.) Portugal is an interesting example of surprisingly rapid market growth of mobile telecommunications in a European country with below-average levels of wealth and economic development. Competition in the sector seems to have been the key factor underlying

---

[60] However, full telecommunications market liberalisation was delayed. Mobile operators had to route international calls over the network of the fixed line monopoly firm until the end of 1999, paying much higher interconnection charges.

this rapid growth. Underestimation of market growth in mobile telecommunications also occurred in many developing countries.[61]

3.4.11 Greece

Greece is the only Western European country that did not have any analogue cellular network. However, the mobile telecommunications sector was liberalised far ahead of the fixed telecommunications market. Greece was the first European country to award licences through a sealed bid auction procedure. The firms Panafon and STET Hellas received GSM licences in 1992. The incumbent fixed line monopolist did not receive any licence at this stage as it was excluded from the bidding. This was exceptional for Western Europe, as the incumbent fixed line monopolist was typically given a GSM licence. Moreover, the licence terms (for which each firm paid $160 million) also included an exclusivity period for all mobile telecommunications frequencies, including for GSM 1800 services, until 2000. Nevertheless, the incumbent fixed line monopolist OTE was able to get *de facto* access to frequencies for DCS 1800 services in 1997, through a presidential decree, and without resorting to any sort of public competition. The ensuing mobile telecommunications firm Cosmote was a joint venture between OTE and the Norwegian firm Telenor. A legal dispute then emerged concerning the exclusivity for mobile services. According to OTE's interpretation of the presidential decree, the company already had the right to provide mobile telecommunications services and this overruled the exclusivity provision for the existing two GSM firms. In this context of legal uncertainty the two GSM firms did not take any extensive legal action. Cosmote launched its services in March 1998 and was very successful in catching up. It has developed the country's only DCS-1800 network, covering more than 99 per cent of the Greek population. Figure 3.16 shows the evolution of the market shares of the three firms. While the market shares of the two entrants move in parallel, with Panafon having a higher market share than STET Hellas, Cosmote's market share quickly exceeded the market shares of the other two. This is the first example where a third entrant is able to become a market share leader in such a quick time.

3.4.12 Central and Eastern Europe

Countries from Central and Eastern Europe (CEE) were late starters in the adoption of mobile telecommunications. The first countries began in the

---

[61] See ITU (1999) for an account of individual cases.

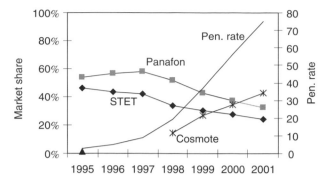

Figure 3.16 *Evolution of market shares of firms and penetration rate, mobile telecommunications market, Greece, 1995–2001*
Note: The market share is based on the total number of subscribers. The penetration rate is calculated as the number of mobile telecommunications subscribers per 100 inhabitants.
Source: Author, using firms' (market shares) and ITU (penetration rate) data.

early 1990s, almost ten years later than Western European countries.[62] All CEE countries chose the same analogue NMT-450 system, while in Western Europe there was a multiplicity of incompatible systems. The great advantages of having a compatible system are economies of scale in the production of equipment and the possibility of 'roaming' – i.e. the user can in principle use the same NMT handset in another country having the NMT system, provided that the firms involved have billing agreements. The mobile telecommunications firms in CEE did not exploit the 'roaming' option, however, and customers could phone only within the home country.

Each CEE country decided to license only one analogue mobile telecommunications firm, which was typically majority-owned by the incumbent fixed line monopolist. Sometimes a foreign minority shareholder was accepted, mainly with the aim of transfer of technological and managerial knowledge. As table 3.6 shows, analogue mobile telecommunications operators charged very high prices, more than double what would have been charged in Western European countries such as the UK. It is therefore no surprise that mobile telecommunications services experienced a slow path of diffusion in CEE. In fact, the penetration rate for a CEE country in 1995 was typically less than one mobile subscriber per 100 inhabitants, while for the UK it was ten mobile subscribers per 100 inhabitants.

[62] For a market description, see Lüngen (1995) and Müller and Callmer (1995).

Table 3.6 *Airtime cost comparisons, 1995*

	Peak	Off peak
Czech Republic	0.81	0.81
Hungary	2.47	1.78
Poland	3.82	0.70
Average	2.37	1.10
UK	1.35	0.54

Note: The figures indicate the cost of a three-minute mobile call in 1995, in US dollars, at purchasing power parity (**PPP**) exchange rates.
Source: Author's calculation from OECD data.

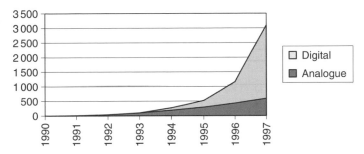

Figure 3.17 *Mobile phone subscribers, Central and Eastern Europe (thousands), by type of technology, 1990–1997*
Source: ITU data.

This situation changed rapidly with the introduction of the digital GSM system. This system was introduced as a regionwide standard and very much in tune with EU Directives. The reasons were many, but in particular these countries expected to become full members of the EU in due course and therefore wanted to adopt regulations and standards aligned to those of the EU. The GSM standard also facilitated European-wide 'roaming'. Since the grant of GSM licences, the penetration of mobile telecommunications has seen a strong acceleration. Figure 3.17 shows the rapid increase in subscribers, broken down by type of technology.

Digital mobile services were also the first telecommunications sector to be opened up to competition. In most CEE countries, competition in digital (GSM) mobile telecommunications was introduced in the form of

Table 3.7 *Starting dates of mobile telecommunications firms, Central and Eastern Europe*

	Analogue	GSM 1	GSM 2	GSM 3
Bulgaria	Dec. 1993	Sep. 1995		
Czech Republic	Sep. 1991	Jul. 1996	Sep. 1996	
Estonia	Jan. 1991	Sep. 1993	Jan. 1995	May 1997
Hungary	Oct. 1990	Mar. 1994	Apr. 1994	
Latvia	Oct. 1991	Jan. 1995	Mar. 1997	
Lithuania	Feb. 1992	Mar. 1995	Oct. 1995	
Poland	Jun. 1992	Sep. 1996	Oct. 1996	
Romania	May 1993	Apr. 1997	Jun. 1997	
Slovak Rep.	Sep. 1991	Jan. 1997	Feb. 1997	
Slovenia	Oct. 1990	Jul. 1996		

Note: The table indicates the status as at the end of 1998.

a duopoly[63] (see table 3.7). The exceptions were Slovenia and Bulgaria, where there was for a long time only one GSM 900 firm. Estonia is remarkable, with three GSM 900 firms. Oligopolistic market structures can have several features. The main distinction is whether the duopoly has simultaneous entry or sequential entry. Latvia, Lithuania and Estonia decided to issue licences sequentially, giving one GSM licence some time ahead of the other. The first entrant typically was the incumbent fixed network operator, which already ran the analogue network. All other CEE countries with two GSM operators instead chose to grant licences at the same time, which meant that the launch date of the networks differed among operators by no more than six months. A further distinction to be made is whether the incumbent fixed telecommunications network firm (and also the analogue mobile telecommunications licence holder) is among the GSM licence holders. Some countries, such as Poland and Romania, did not grant a GSM 900 licence to the incumbent fixed network operator.

The sequencing of entry of the second operators has consequences for the market shares of the firms in question. With sequential entry, one would expect the first entrant to have the larger market share of total subscribers for at least some time. In all three cases of sequential entry in

[63] For a critique of the duopoly model in mobile telecommunications, see McKenzie and Small (1997). The authors show, on the basis of the US market, that the duopoly model induces an inefficient structure, and to redress this problem more firms should be allowed to enter the market.

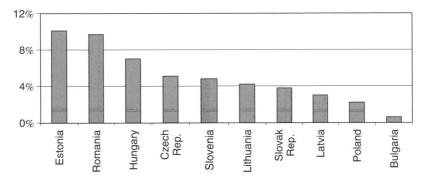

Figure 3.18 *Mobile telecommunications penetration rates, Central and Eastern Europe, 1997*
Note: The penetration rate is calculated as the number of mobile telecommunications subscribers per 100 inhabitants.
Source: ITU data.

CEE (Estonia, Latvia and Lithuania) one could observe a trend towards convergence of market shares. This trend, however, was quite slow. The first entrant generally captured the predominant share of new subscribers, which suggested that early entry gave an advantage in attracting new subscribers. In the case of simultaneous entry one would expect that market shares should be similar. In practice, however, one can observe considerable asymmetries in market shares for some countries (i.e. Hungary, Poland, Romania and Slovakia). The standard deviation of the market shares of the firms with simultaneous entry was 0.112 at the end of 1997, which is only slightly smaller than for sequential entry (0.177). This raises the question of whether the mode of entry has any effect on diffusion at all.

Figure 3.18 shows the penetration rates (mobile telecommunications subscribers per 100 inhabitants) for the CEE countries. It shows that there is a large variation among countries in terms of penetration rates. Country-specific differences in penetration rates may be attributed to several factors – the age of the network, prices charged, competition and prices. Countries which adopted GSM early and who had a competitive framework tended also to have higher penetration rates.

Gruber (2001) has empirically assessed the determinants for the differences in mobile telecommunications diffusion in CEE countries, finding that country-specific factors are strong. This may be related to the fact that mobile telecommunications is a regulated industry and regulatory conditions were highly variable in the CEE countries. The effects of the switch from analogue to digital technology are not significant – a somewhat surprising result as similar estimates for the EU (Gruber and Verboven, 2000) indicated that digital technology was a very powerful factor for speeding up diffusion. Parameters pertaining to competition were of considerable importance, which had a positive and highly significant effect on diffusion speed. Both the entry mode (simultaneous vs. sequential) and the number of firms entering were important. The results provided support for the hypothesis that simultaneous entry speeded up diffusion, and the greater the number of firms, the faster diffusion occurred.

The variables relating to the telecommunications sector itself turned out to be very important, too. The waiting list for fixed telecommunications access had a highly significant and positive effect on the diffusion of mobile telecommunications, corroborating the hypothesis that mobile telecommunications helped to alleviate poor resource allocations in the fixed telecommunications sectors. The fixed telecommunications mainline penetration had a positive impact on diffusion as well. Thus mobile telecommunications were not a substitute for fixed telecommunications, but rather a *complement*. This contrasts with what was found for the EU, where mobile telecommunications was rather a substitute for fixed telecommunications (Gruber and Verboven, 2001a). The fixed line telecommunications penetration rate is much higher in the EU than in the CEE countries. This suggests a hypothesis that at low levels of household penetration of fixed line telecommunications services, mobile telecommunications are complementary, and at high levels they are substitutes. Hamilton (2003) has investigated this question in the context of developing countries. She found that in some cases mobile and fixed telecommunications were substitutes (in particular, when fixed lines networks were non-existent or in a very poor condition) or complements. The latter dominated in most cases. Moreover, mobile telecommunications provided a competitive stimulus to fixed line telecommunications to improve service quality and access.

Gruber (2001) also estimates international convergence in the diffusion of mobile telecommunications for CEE countries. This should be achieved after fifteen years (or after 2005), a result similar to the convergence year (2006) found for the EU (Gruber and Verboven, 2001b). Countries, which entered mobile telecommunications late would thus ultimately catch up because they would have a higher autonomous diffusion speed.

As a policy conclusion for CEE countries it can therefore be said that the emergence of mobile telecommunications has had a very strong impact on the evolution of the telecommunications sector as a whole. Mobile telecommunications not only challenged the natural monopoly paradigm previously predominant in the sector, it also changed the perception of the role of private investment. Anecdotal evidence was already reported that mobile telecommunications had been used as an alternative for the inefficient fixed telecommunications network (Lüngen, 1995). If there were substitution, this would encounter serious technical limits, as the mobile telecommunications network depended heavily on infrastructure owned by fixed line operators. After all, a large share of the calls from mobile phones was to fixed phones, and the usefulness of the mobile phone therefore depended crucially on the efficiency of the fixed network. This is in line with the empirical results found by Gruber (2001), which again suggest that ultimately mobile telecommunications are not an alternative to fixed telecommunications, but rather a complement. Telecommunications sector policies should thus not neglect to further the efficiency of the fixed network.

The lack of easy access to telecommunications services is seen as a bottleneck to economic development. As noted by the World Bank (1994), the large and rising demand for telecommunications services can be met by moving toward a competitive sector structure. Entry of new firms is the single most powerful tool for encouraging telecommunications development because monopolies rarely meet all demand. Mobile telecommunications certainly helps to bring more competition to the telecommunications sector and is also a very effective means of drawing private capital into the sector, especially from abroad. This has been shown to be the case in the CEE countries, as there all mobile telecommunications firms have at least one foreign investor as a shareholder (World Bank, 1998).

3.4.13 The USA

The US cellular mobile telecommunications market is an interesting case study, for several reasons. The cellular system was first developed in the USA, and it is the market where the welfare cost from regulatory delays has been apparently the largest.[64] The regulatory delay also induced the adoption of an inferior technology. In fact, by the time of its adoption AMPS (developed in the 1960s) was less advanced than, for instance, the NMT system, which was developed later.[65] Another important lesson to be drawn from the US market concerns the role of *technology standards*. Analogue cellular telecommunications were introduced in the USA on

[64] Calhoun (1988) describes the sources of the delay in introducing analogue cellular radio, a technology that was conceived in the 1940s, planned in the 1960s and launched in the 1980s. Hausman (1997) has provided econometric estimates of the welfare cost of a regulatory delay of about ten years, caused by regulatory indecision about licensing a monopoly or duopoly and the ensuing lengthy licensing procedure. With welfare costs exceeding $50 billion the estimated amounts turned out to be fairly large.
[65] Calhoun (1998) argues that by the time of adoption the system was technologically already obsolete because it implemented a system fifteen years too late.

the basis of a nationwide standard. This was a great advantage for the diffusion of the technology in a large market such as that of the USA because standardisation reduced the cost of equipment. It also opened the opportunity for nationwide 'roaming'.[66] This technology adoption pattern was in contrast with Europe, where (with the exception of the Nordic countries) there was a variety of (incompatible) systems across countries. This led to a fragmented equipment market and non-exploited economies of scale because the whole economic area of Western Europe could not be uniformly supplied. The possibility of European-wide 'roaming' was also precluded. However, with the switch to the digital technology, the situation was reversed. Europe introduced GSM in a coordinated fashion as a European-wide standard, whereas in the USA the selection of the system was left to the market. In the end, three (mutually incompatible) digital systems established themselves in the market. As already seen in chapter 2, the USA was leading in terms of number of subscribers during the analogue phase, but Western Europe overtook the USA as soon as digital technology was introduced.

Historical context

The concept of cellular mobile telecommunications was developed by the Bell Laboratories.[67] AT&T, the domestic long-distance communications subsidiary of the Bell Group, lobbied the FCC intensively during the 1960s to convince them of the validity of the cellular mobile telecommunications system and its worthiness to be allocated radio frequencies. There was a very lengthy decision process because most of the spectrum was allocated to broadcasting. Eventually the FCC agreed in 1970 to make a tentative allocation of frequencies for cellular mobile telecommunications and invited specific proposals on how to build cellular systems. Bell Labs were the only firm to make a specific proposal by the deadline of end-1971. But only in 1977 was the first authorisation given and trials based on AMPS technology started at the end of 1978. The birth of cellular telecommunications in the USA was therefore held up by regulatory delays. Moreover, when the cellular system eventually came to the market, certain technology bases were already outdated (Calhoun, 1988). For instance, the location of network intelligence with its data processing was still organised centrally at the switching centres, whereas data processing performance had in the meantime developed to a level that favoured decentralised location of network intelligence. As a result, AMPS

---

[66] Though this option was used to only a very limited extent, as AMPS originally did not uniformly organise the ways of transferring the necessary data on the users involved across operators.

[67] For a more detailed historical account, see King and West (2002).

was not as efficient in handling data as some of the competing systems that were introduced at the same time but were of more recent development. AMPS therefore did not take full advantage of the microelectronics revolution that took place at the beginning of the 1970s with the invention of the microprocessor (Cortada, 1987) and which made decentralised data processing feasible.

Licensing

In the light of the successful results from the trial system mentioned above, two other trial licences were granted before the FCC established the rules for the licence allocation in 1982. For reasons of political opportunity to cater for local political pressure groups, the FCC took a very costly route in assigning licences. Instead of giving nationwide licences, the country was divided into 305 Metropolitan Statistical Areas (MSA), defined by a region of at least 100 000 inhabitants, including a town of at least 50 000 inhabitants.[68] For each MSA a duopoly was created for the provision of cellular mobile telecommunications services: in the ensuing 'beauty contest' one licence was earmarked for the local fixed line operator (the so-called 'wireline licence') and one was attributed to new entrants ('non-wireline licence'). The firms had to use AMPS as a standard.[69]

The response to the invitation for bidding was unexpectedly strong. For each non-wireline there was from six to 579 applicants.[70] Even for the wireline licence there was generally more than one bidder, as any of the Regional Bell Operating Companies (RBOC) or local independent telephone company not affiliated with RBOCs could apply. This resulted in the FCC becoming administratively overburdened in assigning all the licences in one step so the selection had to be made in stages. The thirty largest MSAs were dealt with in round 1 and further rounds were organised at five-month intervals. In the successive rounds, the number of bidders increased. For instance, in round 3 (concerning the MSAs ranking from sixty-one to ninety in terms of inhabitants) there were more than sixteen applications per licence. The FCC could not cope administratively with processing such a large number of bids and took advantage of legislation established in 1981 that allowed the FCC to assign spectrum licences for non-broadcasting services by lottery. From round 3 onwards the licences were awarded by lottery; the numbers of participants increased as the pre-qualification criteria for taking part in the lottery were relaxed. It often happened that winners of licences did not have the resources to

---

[68] For details, see Hazlett (1998).   [69] For details, see FCC (1995).
[70] For a more detailed description, see Shew (1994), Parker and Röller (1997) and Garrard (1998).

implement a mobile telecommunications network, and in 1987 the FCC determined that it was legal to trade the licence. The lottery for the thirty licences in round 4 attracted 5182 applications. The subsequent lotteries were organised in six rounds between February and May 1986. There were 92 000 applications for the remaining 185 licences. The FCC also defined 428 Rural Statistical Areas (RSA), with an average population of 150 000. The FCC also adopted a lottery for assigning licences here; however, to avoid opportunistic behaviour, the FCC asked for financial guarantees and a more rapid deployment of the network, reducing the implementation lag from three years to eighteen months.

Overall, it took the USA four years to award cellular licences for its MSAs alone and seven years in total for all A and B licences. This, combined with the delay in deciding on reserving frequency blocks for cellular services at all dissipated the lead that the USA had had over other countries in cellular technology. Both the experience with the administrative burden for assigning licences as well as the fact that an economically valuable public resource such as radio frequencies was being awarded to private agents for free induced the FCC to change the spectrum assignment method during the 1990s, by switching to multiple ascending auctions, as discussed in chapter 6.[71] The spectrum licences for digital PCS mobile telecommunications services were awarded in five blocks in 1988–9 with over 300 000 submissions.

The analogue phase

The period of cellular licensing preceded an important regulatory change in US telecommunications, entailed by the break-up of AT&T at the beginning of 1984. This move established seven RBOCs as completely separate entities from their original parent (Temin, 1987). In line with the original ruling on the separation of cellular business from fixed line telephony, each RBOC was required to establish a mobile subsidiary. There were asymmetries in the opportunities for the bidders; in general, it was far easier for the wireline firms to deploy their networks, because they were more likely to win a licence and could therefore spend more effort in planning ahead. This was also reflected by the network deployment figures. By the end of 1984, twenty-five networks allocated to wireline licences were in operation with only nine from non-wireline licences. This gap increased the following year with eighty wireline licences active and only fifteen non-wireline licences. To ease entry for non-wireline licensees, they were allowed to 'roam' on the networks owned by wireline licence holders

---

[71] Hazlett (1998) also presents political reasons for the switch to auctions as an assignment method.

until their own network was operational. By the end of 1986, most of the ninety largest MSAs had two competing systems.

Because licences were not nationwide but covered only a limited area, users wanting to use the phone in more than one area faced difficulties. Initially there were no 'roaming' agreements among operators from different service areas; there were also technical hurdles due to the fact that the AMPS specification defined only the air interface. Most of the networks were so small that they needed only one switch and there was no standard for transferring data from one switch to another. It was therefore left to the initiative of the firms to define 'roaming' agreements. The wireline firms were in a better position for establishing 'roaming' facilities, because they often had licences for contiguous areas and frequently purchased their equipment from the same supplier, reducing the difficulties of making systems compatible. But even when 'roaming' agreements were in place, they were often difficult to use. There was no automatic recognition of 'guest' users and tedious log-on procedures had to be undertaken. Moreover, to receive incoming calls the caller had to know in which area the user was at the moment. Overall, 'roaming' was very expensive for the user.

These technical difficulties for 'roaming' were tackled over time as networks were upgraded. The consolidation wave in the industry also facilitated the set-up of 'roaming' arrangements. Non-wireline licences were acquired by larger operators such as the RBOCs and entrepreneurs who were more convinced of the long-term potential of the industry than many of the original investors, who wanted only a quick return on their money. For instance, in the top ninety MSAs there were about eighty different licence holders according to the original awards. But by 1992 the top twelve cellular firms, of which the largest was McCaw Cellular, served nearly 60 per cent of the US population. McCaw Cellular had accumulated ninety-one licences for areas with a total residential population of 65 million. This also led to an increase of the value of licences: calculated on a per-inhabitant basis, they rose from \$8 in 1984 to \$270 in 1990 (Garrard, 1998).

The merger and acquisitions (M&A) activity not only had the effect of consolidating the industry; it also blurred the boundary between wireline and non-wireline firms. There was nothing to prevent a wireline firm buying a non-wireline licence in another area where they were not present with a fixed network. At the same time, a number of non-wireline firm bought significant shares in wireline cellular firms. Some operators focused on acquiring licences for contiguous areas; once a number of contiguous licences were held by a single firm, it could start the technical task of transforming multiple, usually incompatible, systems into a single integrated system operating throughout a state or region. To promote 'roaming', the Cellular Telecommunications Industry Association (CTIA) created a committee that set the requirements for the content and format of the data that had to be exchanged for 'roaming' subscribers, setting the basis for the intersystem standard known as IS-41, which was widely adopted throughout the USA.

Transition to the digital phase

The USA were not only late in introducing analogue cellular telecommunications, they were also late in digital systems. There were several reasons for the delayed introduction of digital systems in the USA. Unlike Europe, where the switch to a digital system helped to introduce a standard, the USA had the analogue system (AMPS) already as a national standard. Neither was there the need to push digital technology in order to introduce competition in the sector: with regional duopolies there was already some form of competition. There was thus much less public pressure for change than in Europe, and digital technology was considered from a purely pragmatic point of view. Digital technology was expected to solve the capacity problem in certain metropolitan areas, without requiring any more frequencies.

The FCC decided that cellular firms could introduce new cellular technologies at any time without prior regulatory approval, provided that they were backward-compatible with the existing system. In other words, there was no national digital standard but any new system would have to be backward-compatible with AMPS. The initial specification of the D-AMPS (or IS-54) in fact took very little time to develop, as the change concerned only the air interface.[72] D-AMPS was therefore effectively an interim measure mainly to provide additional capacity. D-AMPS was in fact a dual-mode system; within an existing AMPS system as many channels as needed could be converted to digital. Converting an analogue channel would increase the carrying capacity by a factor of three, and this feature could be used selectively to increase capacity by converting to digital only in congested areas.

It soon became clear, however, that a fully digital system would be needed at some stage to provide a far more comprehensive set of features. IS-136 was supposed to be this final step in the journey to full digitalisation. The evolution was conceived as follows: analogue firms first introduce IS-54 to relieve capacity problems and then, once the customers need the full facilities of a digital system, IS-136. Some major mobile

---

[72] This was achieved by switching to TDMA.

telecommunications operators such as AT&T Wireless, Southwestern Bell and Bell South, implemented these plans for D-AMPS, which was often referred to as TDMA. But the journey to a digital system was ultimately not as smooth as intended. Alternative technical solutions were proposed and accepted, with the consequence of abandoning the idea of a standard altogether.

The first alternative was a narrow-band analogue solution, usually referred to as N-AMPS and proposed by the equipment supplier Motorola. The width of AMPS channels would be reduced from 30 kHz to 10 kHz, allowing three times as many channels in the same amount of spectrum. This solution would tackle only the capacity problem and leave the introduction of digital systems to a later stage. Eventually N-AMPS was rejected by the CTIA.[73]

A more successful alternative emerging was CDMA, promoted by the US firm Qualcomm. In 1994, CDMA was adopted as a second cellular system in the USA under the name IS-95. Qualcomm set up a CDMA development group to promote the system more widely. All major equipment manufacturers joined this group, with the notable exception of Ericsson, which maintained that CDMA would have no advantages over GSM and it did not intend to manufacture any products for this technology.[74]

Because only the busiest cells were converted to TDMA or CDMA, mobile handsets had to become dual-mode (i.e. AMPS + TDMA or AMPS + CDMA). Digital dual-mode handsets were not only larger than their analogue counterparts, they also cost more than twice the analogue version, so only less than 10 per cent of subscribers had a dual handset in 1994. The immaturity of CDMA technology, and the limited availability of terminals, delayed the introduction of the technology by those firms who had adopted CDMA. Airtouch was the first firm to do so in mid-1996 (i.e. four years later than TDMA). 'Roaming' between one network with CDMA and another with TDMA is possible only using analogue channels, the only common factor that remains between the firms. As long as the FCC requirement for nationwide compatibility remained in force, a complete change to digital services appeared impossible from a regulatory point of view. By 1995, about 60 per cent of mobile telecommunications users were customers with wireline firms and 40 per cent with non-wireline firms. In 1995, the top five mobile telecommunications firms accounted for 63 per cent of subscribers. The market thus resembles an oligopoly. It is

---

[73] However, N-AMPS was adopted in countries such as Venezuela, Guatemala and the Philippines.
[74] Ericsson eventually had to renege on this strategy with 3G mobile technology, when it adopted an updated version of CDMA technology (W-CDMA) developed by Qualcomm.

therefore not surprising that the FCC (1995) reported higher than average profitability for the industry, in particular for metropolitan areas. These findings of limited competition reinforced the view that the industry should accept additional entry.

Personal communication system (PCS)

It became clear during the second half of the 1980s that cellular mobile telecommunications would mainly cater for the business market. The USA made similar efforts as in Europe (in particular, in the UK) to create an application of mobile telecommunications services that would be for the mass market, an objective that was deemed not feasible with existing cellular networks. In the UK these propositions were referred to as 'personal communications networks' (PCNs), in the USA they fell under the heading of 'personal communication services' (PCS). While the UK was able to grant three licences for PCN services within the same year that the proposals were made (1989), in the USA it took six years until the first PCS licences were granted in 1995. Regulatory delays were again the main culprit for the late introduction of new services.

The FCC made several changes in the spectrum allocation procedures. Instead of using the same MSA and RSA definitions adopted for the original AMPS licences, the FCC proposed using fifty-one larger units know as major trading areas (MTAs), subdivided into 493 basic trading areas (BTAs). For each MTA there would be two licences (A and B licences) with 2×15 MHz each and one 2×15 Mhz licence for each BTA. Moreover, each BTA would also have three licences (D, E, F licences) with a smaller frequency band of 2×5 Mhz. Overall, any town in the USA should be in principle covered by up to six PCS licences. MTA licencees were required to cover 67 per cent of the population after ten years, and BTA licences obliged firms to cover 25 per cent of the population after five years.

The FCC also changed the allocation method, opting for auctions. The MTA auctions started in December 1994 and lasted until March 1995, yielding a total of $7.8 billion in licence fees. As expected, the market structure was much more concentrated from the beginning than with respect to cellular licences. Large firms conquered most of the licences: established operators, along with Sprint's mobile subsidiary WirelessCo, won thirty out of the ninety-nine available[75] licences (see table 3.8). Overall, the first three firms received almost two-thirds of the total licences granted.

---

[75] Three so-called 'pioneering licences' had already been awarded previously.

Table 3.8 *Largest holders of MTA (A and B block) licences, USA, 1995*

Firm	Number of MTA licences
Sprint Spectrum (WirelessCo)	30
AT&T Wireless PCS	21
PrimeCo Personal Communications	11
Other	37
Total	99

Source: FCC data.

There was an escalation of interest in C licences, originally intended as an encouragement to local entrepreneurs. As each individual BTA would cover on average one-tenth of the population of a MTA, it would be much harder for any firm to obtain wide areas of coverage. Several 'affirmative action' provisions were originally made for the licensing process: for instance, 'designated entities' such as small firms owned by women or minorities would qualify for a 25 per cent discount on the bid price and could spread payments over six years. After a legal challenge followed by a Supreme Court ruling, the references to minorities and women eventually had to be dropped and only preferential treatment for small firms was upheld. The auction started in December 1995 and ended in May 1996. After 184 rounds of bidding a total licence fee of $10 billion was raised. This means that the price of a licence per inhabitant of the area the licence refers to was much higher than for the MTA licences.

This escalation of licence fees as a result of the auction process also led to the first firm failures. The firm BDPCS failed to pay for the seventeen licences it had won, as did National Telecom for its own licence. BDPCS had an aggressive bidding strategy speculating on the fact that winning a licence would attract the backing from a large firm without a licence. However, this possibility did not materialise. The eighteen licences were re-auctioned in July 1996. At the end, 493 licences were distributed to about eighty firms. Thus the market structure was quite fragmented, with the first three firms accounting for one-third of the total licences granted. NextWave was the largest, obtaining sixty-three licences (see table 3.9).

The auction for the remaining 1479 narrow-band licences (D, E and F block) started in August 1996 and ended in January 1997. Any firm could bid in these auctions, including existing mobile telecommunications firms. The auction raised a total of $2.5 billion, which is less than the A, B and C block if calculated per MHz. This reflects the perceived lower utility of so-called narrow-band services.

Table 3.9 *Largest holders of BTA (C block) licences, USA, 1996*

Firm	Number of BTA licences
NextWave Personal Communications	63
Pocket Communications	43
Omnipoint PCS	18
Other	369

Source: FCC data.

As with digital cellular systems, the FCC did not set any national technical standards for PCS (apart from those essential for interference and safety). IS-54 (TDMA) and IS-95 (CDMA), the two digital systems already in place for cellular services at 800 MHz, provided the natural platforms for PCS voice services. Just like the GSM 1800 services in Europe, the system needed an upgrading for operation at twice the frequency of the original specification. However the firm APC, which was granted a trial licence to develop PCS services, opted for an adaptation to 1900 MHz of the European GSM 1800 system based on GSM technology. From a technological point of view, this was a low-risk strategy, as GSM 1800 systems had already been implemented in the UK and Germany. Moreover, unlike US cellular firms, APC did not have to cope with an installed subscriber base and hence backward compatibility was not an issue.

Finally, three digital cellular mobile telecommunications systems found adoption in the US market. They essentially were IS-136 (D-AMPS) and CDMA as digital systems developed for the USA market, as well as a version of GSM adapted to the prevailing frequencies allocated in the USA for mobile telecommunications, known as PCS 1900, or IS-661, or GSM 1900.[76] Table 3.10 shows the breakdown of MTAs by technology choices and population size of areas licensed. CDMA attracted the largest number of MTA and populations. TDMA counted on more MTAs than PCS 1900, but on fewer population. The reason for this split market was that none of the systems represented a superior choice in each respect. CDMA was believed to be the superior

---

[76] PACS was a fourth cellular mobile telecommunications system in the US market, developed by Bellcore about ten years before. It was particularly suitable for fixed local loop applications. As a PCS system it was, however, adopted only in Alaska, which is generally not reported in statistics. Likewise, there is also a system called 'integrated digital enhanced system' (also known under the heading iDEN). This is a digital dispatch radio system that also has features resembling those of a cellular mobile telecommunications service system. The firm Nextel is supplying these services in the USA, and has found fairly wide diffusion.

Table 3.10 *US PCS network coverage, by technology, 1996*

	CDMA	TDMA	GSM 1900
Number of MTAs	46	33	21
Coverage area population (million)	243	114	140

Source: FCC data.

Table 3.11 *Mobile telecommunications systems adopted by different licence holders, selected MTAs*

Frequency range licence type	800 MHz		1900 MHz		
	A	B	A	B	C^a
New York	TDMA	CDMA	CDMA	GSM	CDMA
LA/San Diego	CDMA	TDMA	GSM	CDMA	CDMA
Chicago	CDMA	TDMA	CDMA	TDMA	GSM
Washington/Baltimore	CDMA	TDMA	GSM	TDMA	CDMA

Note: [a]Central main city only.
Source: FCC data.

system in terms of providing subscriber capacity, but it was not entirely ready from a technological development point of view. TDMA had the advantage of backward compatibility with the already established D-AMPS networks, but as a technological choice it was considered as inferior with high terminal equipment prices.

The US digital mobile telecommunications market was therefore faced with a multiplicity of incompatible systems. Table 3.11 illustrates the technology adopted by the different licence holders for some selected MTAs. Whereas in Europe there is a standard for all countries, a US consumer, depending on the area, may have to face up to three different incompatible systems. This wide variety of mobile systems is likely to create confusion and uncertainty in the user.

In spite of the considerable emphasis on the fact that PCS embraced a wide range of potential services, the ultimate aim of the licence winner turned out to be to deliver straightforward mobile telephony in competition with all the established cellular firms. As in Europe with GSM 1800, the services of PCS firms in fact became indistinguishable from those provided by cellular firms operating networks at the 800 MHz frequency range.

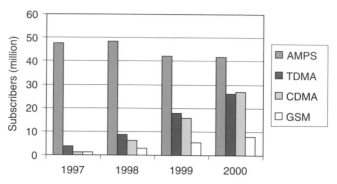

Figure 3.19 *Mobile telecommunications subscribers, by technologies, USA, 1997–2000*
Source: FCC data.

Figure 3.19 shows the subscriber evolution of the three digital technologies over time and compares them with the analogue technology AMPS. The digital technologies pick up slowly and late when compared with Western Europe. Within the digital technologies, TDMA has a head start, but CDMA has overtaken TDMA by 2000. The analogue subscriber base is expected to persist for quite some time because of the FCC decision that obliges firms operating networks in the 800 MHz range to provide AMPS services at least until 2007 (FCC, 2003). The extent to which competition in technological systems has held back subscriber growth will be analysed in chapter 4.

By the end of 2002, the mobile penetration rate stood at 49 subscribers per 100 inhabitants, which is low if compared to Europe or Japan. This may appear as puzzling, considering that indicators such as number of firms or level of prices suggest that competition should be intense. Table 3.12 shows that the revenue per minute of use is typically much lower than in many other countries with higher penetration rates. The minutes of use are also much higher. Moreover, in most of the US regions there are seven firms in the market, whereas in Europe there are typically three–four firms. The FCC (2003) indicates the pricing regime as the main reason for the low penetration rate in the USA. Only the USA and Canada have the receiving (or mobile) party pays (RPP) regime, by which the mobile subscriber pays for any incoming call, with or without 'roaming'. All other countries have a calling party pays (CPP) regime (i.e. the mobile user does not pay for incoming calls, unless she is 'roaming'). This issue will be taken up further in chapter 5.

Table 3.12 *International comparisons of performance indicators, 2002*

Country	Pricing regime[a]	Penetration rate[b]	Usage (minutes)[c]	Airtime revenue[d]($)
Canada	RPP	37	270	0.11
USA	RPP	49	458	0.12
Japan	CPP	62	170	0.30
France	CPP	63	156	0.20
South Korea	CPP	68	296	0.10
Australia	CPP	68	173	0.16
Germany	CPP	72	72	0.29
UK	CPP	85	132	0.22
Finland	CPP	85	146	0.24

Notes: [a]CPP = Calling party pays; RPP = Receiving party pays.
[b]Penetration rate is the number of mobile subscribers per 100 inhabitants.
[c]Usage is the number of outgoing traffic minutes per month and per subscriber.
[d]Airtime revenue is the revenue per minute of use.
Source: FCC data.

3.4.14 Japan

The Japanese cellular mobile telecommunications market, like that of Germany, France and Italy, illustrates the evolution of a national standard within a monopoly framework. The impetus for market growth emerged only from the introduction of competition, which in this case came mainly through outside pressure from US trading interests.[77] The Japanese telecommunications monopoly NTT was the first firm worldwide to run a cellular network in 1979. NTT developed and used its own technical standard, but the early adoption of cellular mobile telecommunications did not facilitate fast market growth. The network started in Tokyo first, and nationwide coverage was available five years later. Figure 3.20 compares the penetration rates (subscribers/100 inhabitants) in Japan with other European countries where there was an analogue monopoly with a national standard. It shows that after ten years of operation a penetration rate of fewer than 0.7 subscribers/100 inhabitants was in line with most of the European monopolies, but far below what was experienced in duopoly markets such as the USA (and UK).

---

[77] Accounts such as by Prestowitz (1988) well reflect the spirit of that period of trade US–Japan disputes.

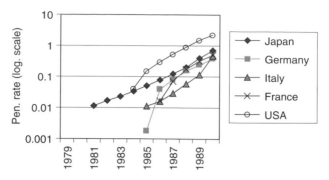

Figure 3.20 *International comparison of penetration rates, early phases of the mobile telecommunications market, 1979–1989*
Note: The penetration rate is calculated as the number of mobile telecommunications subscribers per 100 inhabitants.
Source: ITU data.

The early launch of the cellular system thus seems to have been driven more by technical curiosity rather than serious commercial intent (Garrard, 1998). NTT had a monopoly position in the market as most European telecommunications operators and behaved in a similar way: prices for equipment and services were very high, choice of terminals was limited and customer care was poor. Figure 3.21 shows the evolution of mobile telecommunications subscribers in Japan and indicates an increase in market growth towards the end of the 1980s. The market situation improved when in 1986 two new cellular licences were granted: one to Nippon Ido Tsuschin (IDO) for the Tokyo–Nagoya region and one to Daini Denden Inc. (DDI) for the rest of the country. A duopoly market regime was thus established for each region, with IDO using an NTT technology-based system and DDI using a Japanese version of TACS ('JTACS'). There was thus a duopoly with the coexistence of two different technologies, a situation similar to France. As in France, the coexistence of different technologies in the market seems to have inhibited growth in Japan, too. The duopoly led to accelerated subscriber growth with subscribers doubling in 1989, and again in 1990. Even though the competition mechanism seemed to work, the conditions imposed on new entrants limited its effectiveness. Under the terms of the licences, the network operators were forced to continue renting terminals, keeping prices high and limiting the benefits of network competition. Though the tariffs charged were substantially lower than NTT's, the absolute level still was extraordinarily high.

Figure 3.22 shows the comparative evolution of mobile telecommunications penetration rates in Japan, the USA and the EU. The penetration

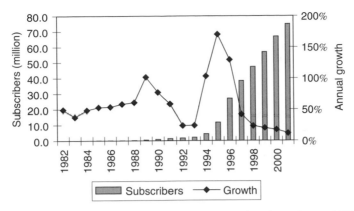

Figure 3.21 *Evolution of mobile telecommunications subscribers, Japan, 1982–2001*
Source: ITU data.

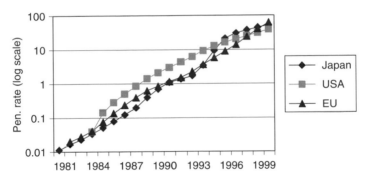

Figure 3.22 *Evolution of mobile telecommunications penetration rates, Japan, the USA and the EU, 1981–2000*
Note: The penetration rate is calculated as the number of mobile telecommunications subscribers per 100 inhabitants.
Source: ITU data.

rate is indicated on a logarithmic scale, showing that in spite of the duopoly starting in the second half of the 1980s overall mobile penetration in Japan was still at relatively low levels. Only with the further introduction of firms did the penetration rate rise and during the second half of the 1990s Japan leapfrogged the USA and Europe. By 2000, after a merger wave, Japan fell behind, with Europe taking the lead in terms of penetration rate.

Coming back to the analogue period, it seems that also Japan suffered from the competition of *incompatible systems*. The use of incompatible systems in different areas of the country not only restrained market growth but also became the subject of the trade dispute between the USA and Japan already mentioned.[78] Motorola, which supplied equipment to DDI, claimed that it had been promised that spectrum for the JTACS system would be allocated in the most attractive areas such as Tokyo and Nagoya, to allow DDI users to 'roam' onto networks in these key business areas. The necessary bandwidth was made available only as a result of considerable pressure from the US government and even more significantly IDO, which launched its own JTACS network at the end of 1991. As the disadvantages of incompatible systems became evident, the Ministry of Post and Telecommunications organised the Japanese Digital Cellular Radio System Committee in 1989, with a brief to define the technical requirements of a national system. The core characteristics were defined by the end of 1990, borrowing many parameters from D-AMPS. In support of market opening measures, foreign equipment suppliers such as Motorola, Ericsson and AT&T were invited to join Japanese companies in the development of practical systems. Like IS-54 in the USA, the specifications of JDC covered only air interface and not the entire system, which became manufacturer-specific.

Japan was also planning to create a PCS. In 1991, the government granted two additional licences to the firms Tu-Ka and Digital Phone for the provision of digital cellular services in the 1500 MHz range. They competed in densely populated areas, while their JV firm Digital Tu-Ka covered less densely populated areas. NTT, IDO and DDI all operated digital systems at 800 MHz, but only in their franchise area for analogue cellular services.

NTT's mobile telecommunications subsidiary NTT DoCoMo launched the first digital system (PDC) in 1993, but the effect on the market was limited. Things changed in 1994 when Tu-Ka and Digital Phone launched their systems as well, followed by IDO and DDI. The market grew strongly after the entry of these additional competitors. Both digitalisation and additional entry allowed Japan eventually to overtake the USA in terms of penetration rates (see figure 3.22). Towards the end of the 1990s a merger wave occurred. In 1999, Digital Phone and Tu-Ka merged, forming J-Phone, which became part of the Vodafone group. In 2000, IDO and DDI merged to form the new firm KDDI. KDDI gained access to the 1800 MHz spectrum as it also took over the overlapping network elements of the constituent parts of J-Phone. The market structure is highly asymmetric, with NTT DoCoMo accounting for 59 per cent of subscribers in 2001. The second largest firm is KDDI with 24 per cent of subscribers,

---

[78] See Tyson (1992) for a description of the trade dispute over mobile telecommunications equipment.

followed by J-Phone with 17 per cent. In principle, all three firms have digital networks based on PDC. But on the 800 MHz frequencies, KDDI also started to change technology, replacing PDC with CDMA. Thus, while the Japanese market had a growth rebound during the 1990s, with digitalisation, standardisation and entry of new firms, the country fell back again at the end of the decade following increased market concentration and the introduction of incompatible systems.

3.4.15 Australia

The Australian mobile telecommunications market is interesting because it displays frequent change in technologies.[79] Cellular mobile telecommunications started with an AMPS network in 1986, operated by the incumbent telecommunications firm Telstra. In spite of coming late to the market and with a monopoly regime, the Australian market grew rapidly, reaching a penetration rate of 5 per cent by 1993, which put it among the highest in the world. The subscriber growth rate increased with the entry of Optus, which also operated an AMPS network.

Digital mobile telecommunications services were introduced in 1994 by the two incumbent analogue network firms plus Vodafone as new entrant. The country thus abandoned the adoption path of D-AMPS, used in most countries with AMPS networks. Further radio spectrum in the 1800 MHz range was made available in 1998, allowing the entry of three more firms, One.Tel, Hutchison and AAPT. Out of the new entrants, only One.Tel adopted GSM, while the other two firms opted for CDMA. Among the incumbent firms, Telstra adopted a CDMA 800 network in parallel to its GSM network, with the aim of gradually moving the customers it had on the AMPS network to CDMA. The AMPS network was closed in 1998. Of the new entrants AAPT started to build the network, but actually never launched the service. So there were only five firms supplying services in 2000. The technology mix present in the Australian market is indicated in table 3.13. Four firms out of five supply digital mobile telecommunications services using GSM, and two firms use CDMA.

Figure 3.23 shows the evolution of market shares in the mobile telecommunications market in Australia. An asymmetric pattern is emerging. Telstra had a declining market share until 2000, following the entry of the rival firms. However, since then market shares appear to have stabilised. Optus had a much larger market share than Vodafone, reflecting the fact that the first had a nationwide network, whereas the second was focusing on major urban

---

[79] For more background information on the Australian market, see Garrard (1998) and ACA (2000).

Table 3.13 *Technology choices by firms, Australian market, 2000*

Firm	AMPS	GSM	CDMA
Telstra	x	x	x
Optus	x	x	
Vodafone		x	
Hutchison			x
One.Tel		x	

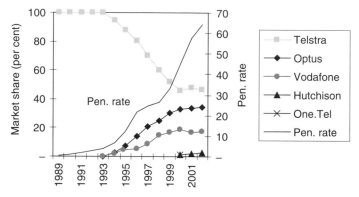

Figure 3.23 *Evolution of market shares and penetration rates, mobile telecommunications market, Australia, 1989–2002*
Note: The market share is based on the total number of subscribers. The penetration rate is calculated as the number of mobile telecommunications subscribers per 100 inhabitants.
Source: ACA data.

areas. Overall, the market share ranking reflected the entry pattern, with earlier entrants having larger market share patterns than later entrants.

3.4.16 South Korea

South Korea had a long period of analogue monopoly by SK Telecom, based on the AMPS standard. Subscriber numbers developed very sluggishly and by 1993 the country had a mobile penetration rate of 1.1 subscribers per 100 inhabitants.[80] At the beginning of the 1990s the government wanted to put out to tender a second licence for analogue mobile

---

[80] An account of the analogue phase can be found in Garrard (1998).

telecommunications services. However, the final assignment was with-drawn from the winner amid accusations of political cronyism. South Korea was relatively late in taking the decision to introduce digital mobile telecommunications services. There was an initial intention to adopt D-AMPS, but in 1993 the government announced that the standard to be adopted would rather be CDMA. South Korea thus became the first country to adopt the CDMA system worldwide. Behind this decision was clearly a strategy to promote the local telecommunications equipment industry. CDMA technology was commercialised through a joint effort by the Electronic Telecommunications Research Institute, the South Korean equipment manufacturer Samsung and the US equipment manu-facturer and key CDMA patent holder Qualcomm. In this sense, South Korea wanted to replicate the success European equipment manufacturers had with the introduction of GSM.

Market development was promoted through massive entry. Two licences for digital mobile telecommunications services based on CDMA technology were assigned in 1995: one to SK Telecom and one to Shinsegi Telecom. At the same time, three PCS licences, based on CDMA technology in the 1800 MHz range, were awarded to Korea Telecom Freetel (KTF), LG Telecom and Honsal. The entry of four new firms and the establishment of a nationwide standard led to a very rapid expansion in the number of subscribers until 2000. Concomitant with this slowdown was a consolida-tion wave in the industry, reducing the number of firms from five to three. SK Telecom merged with Shinsegi and KTF with Honsal. The first merger was particularly contentious, because SK Telecom had already a market share of 42 per cent before the merger, which afterwards increased to 53 per cent. KTF was the second largest firm, with 32 per cent of subscribers, followed by LG Telecom, with 15 per cent. The merger was approved in 2002 with the stipulation that the joint market share should not exceed 50 per cent.[81] SK Telecom was in a situation where it had to reduce subscribers, which did not happen in 2003 as it increased market share. Figure 3.24 illustrates the evolution of per cent of market shares in the mobile telecommunica-tions market in South Korea. During the first three years after entry the incumbent SK Telecom lost market share, but then stabilised around 40 per cent of subscribers. This illustrates a highly asymmetric market share, with a return to increasing market dominance apparently emerging after the merger wave.

After 1999, the Korean market showed a reduction in the growth rate of subscribers. Figure 3.25 shows the evolution of the penetration rate in the Korean market and compares it with Japan and Australia. These countries

---

[81] See ITU (2003b) for more details.

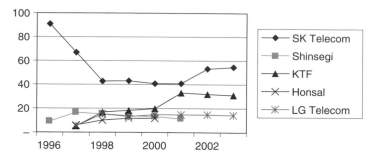

Figure 3.24 *Evolution of market shares, mobile telecommunications market, South Korea, 1996–2003*
Note: The market share, in per cent, is based on the number of subscribers.
Source: Author using ITU data.

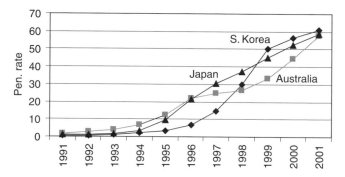

Figure 3.25 *Evolution of penetration rates, South Korea, Japan and Australia, 1991–2001*
Note: The penetration rate is calculated as the number of mobile telecommunications subscribers per 100 inhabitants.
Source: ITU data.

leapfrogged each other during the 1990s. At the beginning of the 1990s Australia was leader, having AMPS with a competitive supply. By the mid-1990s Japan was taking the lead as PDC was introduced with a competitive supply. Korea leapt in front from 1999 onward with the introduction of CDMA and massive entry of new firms. In all of these cases the prevalence of a national standard along with new entry seems be an essential ingredient for rapid market growth.

3.5 Conclusion

This chapter has made an extensive survey of the evolution of national markets for cellular mobile telecommunications services. Entry by new firms and the choice of technological systems have received special emphasis. From these country-specific cases some common features emerge. A uniform standard seems favourably to affect growth in terms of mobile telecommunications subscribers. This effect is enhanced if there is competitive supply of services. This is particularly striking when comparing the US market with the Western European national markets in the analogue and digital phase. In the analogue phase, the USA had a standard, whereas in Europe this was not the case: the US market grew much faster than the European market. This situation was reversed with the switch to the digital phase. Europe introduced a continent-wide standard, whereas the USA allowed competition among multiple, incompatible systems: Europe saw a much faster diffusion of digital mobile telecommunications services. The mobile telecommunications market seems to provide support for the hypothesis that competition on a standard drives market growth more than competition among systems. However, competition among systems seems to provide more incentives for technological advance. A trade-off between *static* and *dynamic efficiency* thus seems to emerge. The chapter also looked at some more peripheral countries or regions, where these considerations appear to apply as well. The aim of chapter 4 is to move on from this country-specific evidence and to see whether the propositions made on market growth are also more generally valid. Statistical evidence is presented using all adopting countries in the sample under investigation.

4 The determinants of the diffusion of cellular mobile telecommunications services

4.1 Introduction

This chapter, based on Gruber and Verboven (2001b), provides an econometric analysis of the determinants of the diffusion of cellular mobile telecommunications. Whereas in chapter 3 the emphasis was on a description of the idiosyncrasies of national markets in the evolution of the mobile telecommunications market, the aim is now to find some statistical trends concerning the driving forces for market growth. Several econometric studies have considered the cellular mobile industry in an individual country or a restricted number of countries, focusing on market conduct (Ruiz, 1995; Parker and Röller, 1997; Nattermann, 1999), or on the role of country characteristics for diffusion (Dekimpe, Parker and Sarvary 1998). The hypothesis that firms in this industry have market power is well supported by the data, but this has not yet been linked up with the diffusion literature This chapter illustrates the role of regulatory decisions, in particular on entry and setting of standards, and other market parameters for the diffusion of cellular mobile telecommunications. The aim of this chapter is therefore to unravel the effects of entry regulation and standard setting on the evolution of a specific industry, that of worldwide cellular mobile telecommunications services.

The chapter is organised as follows. Section 4.2 provides a brief survey of the issues and presents some diffusion models. Section 4.3 describes the econometric model. Section 4.4 presents and discusses the empirical results. Section 4.5 draws some brief conclusions and discusses implications for public policy.

4.2 Preliminary considerations on diffusion and market structure

4.2.1 Survey of the literature

There has been significant research on the adoption of new technologies, from both theoretical and empirical points of view.[1] A general theme that

---

[1] For a survey of the theoretical literature, see Stoneman (1983) and Reinganum (1989). For more empirically oriented descriptions, see Ray (1984), Rogers (1995) and Mowery and Rosenberg (1998).

emerges from this work is the striking variation in the speed of introduction and diffusion across sectors. In the semiconductor industry, for example, the adoption of new technologies was extremely rapid (Gruber, 1994). In other sectors, innovations spread at a much slower pace, many years after the technological innovation became first available (Ray, 1984). Diffusion patterns thus varied widely across both sectors and countries. For instance, Gruber (1998) has shown for the textile industry that shuttleless looms diffuse rather slowly, but both country-specific variables such as wage rates and overall policy variables such as trade liberalisation matters. This study has confirmed the theoretical hypothesis emerging from traditional diffusion models that increased competition tends to speed up the diffusion of innovation,[2] but there are also studies for industries where this is not the case.[3] In telecommunications, it has been argued that competition creates additional incentives to reduce costs, to innovate and to eliminate distorted prices (Laffont and Tirole, 2000). While there has been some empirical work on the role of market structure and competition in the diffusion of innovation, the effects of the timing of entry have not been systematically considered. Important empirical issues on entry in telecommunications are the impact of regulatory delay in issuing first-entry licences on the diffusion of innovation; the pre-emptive, immediate and long-term effects of additional entry licences on the diffusion of innovation; and the distinction between simultaneous and sequential entry. The evidence for the role of compatibility of systems (standardisation) in the diffusion of innovations has yielded only mixed empirical evidence. For instance, the failure in setting a standard may lead to slow diffusion because of the reluctance of adopters to take the risk of investing in technologies which may not succeed. But there is also the contrary risk of government setting standards that may be inefficient and thus inducing inferior adoption patterns.[4] In spite of the significant research in this area, the determinants of diffusion still remain only partially understood.

Mobile telecommunications services provide an interesting case on how well identified events, such as technological switches and regulatory changes, have affected the diffusion pattern. As seen in chapter 2, the technology of wireless communications has been available a long time. However, only after basic innovations in semiconductor technology (such as the microprocessor) did the adoption of the wireless telecommunications became feasible on a

[2] Other studies coming to similar conclusion are, for instance, Oster (1982) and Levin, Levin and Meisel (1987)
[3] See for instance Rose and Joskow (1990).
[4] These issues of excess inertia and excess momentum are extensively discussed in Shapiro and Varian (1999) and Rohlfs (2001).

large scale. As mobile telecommunications is a network industry there are so-called 'externalities' that affect the diffusion. Some determinants of diffusion may also be subject to policy decision. Because of the rapid technological changes in this industry, it may be difficult to establish consensus on the optimal policies to be followed. Important decisions on market structure have to be made, such as the number of firms and their timing of entry. A further issue is the public interest for setting a technological standard: should a standard be mandatory, or should the market through competition among systems establish a standard? Another way of putting this question is: do standards create markets or markets create standards? Apart from unresolved theoretical issues, there has been little empirical work on the effects of public policy decisions on the diffusion of new technologies.

The theoretical literature on technological standards, however, seems to converge on some issues. With positive network externalities (for example, when consumers value a system more the more users adopt it), for instance, standards lead to faster market growth. Moreover, standards tend to benefit consumers as they reduce their search and switching costs. On the supply side, standards reduce the scope for product differentiation. Price competition may thus be enhanced because firms have to compete using the same standard. But there is also the risk that a selected standard will not be the most efficient one: because of lock-in effects, it may become difficult to switch to a better one. Nevertheless, there are several industries where different incompatible systems coexist and other cases where market forces push one system to take the whole market, establishing itself as the standard (e.g. the VHS system for video recorders). Definite answers on market outcomes and social welfare implications crucially depend on the market and technology parameters involved. Despite the extensive theoretical literature, there exists no empirical work that compares the effect of imposing standards on the diffusion of a new technology with the effect of allowing multiple systems to compete.[5] Network effects typically lead to 'tipping' markets, where the winning technology takes the whole. Advocates of free markets point out that governments trying to influence this game risk promoting inferior standards.[6] The counterargument is that markets may lock themselves into inferior outcomes and government intervention may then become necessary to cope with this network externality.[7] In the case of cellular telecommunications, government intervention can help to promote

[5] For an analysis of the presence of network effects, see Saloner and Shepard (1995). They do not directly compare competing systems with single standards.

[6] An example is high-definition television in Japan, where the government promoted an analogue standard, neglecting the fact that the worldwide evolution was toward digital technology (Shapiro and Varian, 1999).

[7] The typical example reported in the literature is the QWERTY keyboard winning over the allegedly superior Dvorak keyboard. But Liebowitz and Margolis (1999) confute these

national systems internationally (as the examples of NMT, AMPS and GSM show). But from an efficiency point of view (such as spectral efficiency) the systems that get established may not be the best ones, as the case of NMT shows (this holds to some extent also for GSM). But in none of the cases have the systems cornered the market fully, neither on an international nor on a national level, when there was competition among systems (e.g. the US digital cellular market supports three systems). Because of the obligations of interconnection among networks, Shapiro and Varian (1999) argue that the cellular mobile telecommunications market, in spite of strong network effects, may not be particularly prone to tipping.

The advantages of a single standard specifically for cellular mobile telecommunications derive from the fact that it overcomes the problem that various components of the system are incompatible with each other. When there are competing systems, a firm needs to make infrastructure investments that are specific to the technological system used. It also means that mobile users can use their handset (mobile phone) only within the areas that support their system. This creates network externalities in various ways. Consumers who use their handset only near their homes would prefer that competing operators offered the same technological system, since this would allow them to switch without a need to buy a new handset. Yet such consumers would not care whether competing operators in other areas also offered the same system. In contrast, consumers who 'roam' across the country would gain from having a single, nationwide system. This would allow them to use their handset wherever they were located. Depending on the mobility of consumers, network externalities are thus local, national, or even international in scope. In addition to reducing switching costs and creating 'roaming' possibilities, the presence of a single technological system also has the advantage of exploiting economies of scale in the manufacture of equipment. A main disadvantage of having a single standard is that new technological systems have little chance to succeed, even if they are of a better quality. This argument has been made at several places, particularly in the US debate during the digital technology phase, where multiple systems competed.

4.2.2 Basic regulatory decisions in the mobile telecommunications industry

As argued already in chapter 3, the cellular mobile telecommunications industry offers an interesting opportunity to make a comparative analysis,

---

arguments after careful investigation of the case and claim more generally that the set of theoretically possible and empirically relevant examples of market failures in picking standards is in fact empty.

since countries have followed quite varied and changing policies regarding both entry regulation and standard setting. Regulatory intervention in mobile telecommunications involves several dimensions. The focus is now on the decisions affecting the industry before services are actually supplied. First, the government needs to decide whether to set a single national (or international) standard, or whether to allow multiple technological systems to compete. Second, the government has to decide how many operators will be granted licences. (This also involves an important decision with respect to the timing of first and additional licences.) Third, the government needs to decide how to grant licences. In the early days of mobile telecommunications, licences were often granted on a first-come-first-served basis. With the introduction of cellular technology, the first licences were frequently granted by default to the incumbent fixed operators. Additional licences were granted either through an auction, or through an administrative tender procedure (or 'beauty contest'), possibly including a licence fee. In this econometric study the focus is on the first and the second dimensions of the licensing policies: a single standard vs. competing systems and timing of first and additional licences. The room for discretionary policy is limited by the available spectrum capacity and by the technological options. The policy decisions are described by the 2×2 policy matrix in table 3.1 in chapter 3 (p. 68). The columns indicate the decision between a single standard or multiple competing systems. The rows refer to the decision to admit a monopoly or competing firm.

In mobile telecommunications firms may seek to extend market power by raising switching costs for customers. Switching costs are therefore a potential determinant of competition effects in the mobile industry. Mobile operators frequently offer long-term contracts to consumers, for example, thereby artificially creating lock-in.[8] In a one-period context, switching costs (like product differentiation) tend to soften competition between operators. In a dynamic setting, switching costs may induce firms to compete more aggressively for market share during the early phases of competition. The presence of switching costs gives rise to some testable predictions. First, switching costs can explain how the timing of competition affects the diffusion of innovation. When entry is simultaneous, operators obtain more or less symmetric market shares, allowing them to compete rather 'softly'. In contrast, when entry is sequential, the entrant has to compete aggressively to obtain customers from the installed market share of the incumbent firm.[9]

---

[8] As mentioned already, switching costs are further enhanced when there are multiple systems, forcing consumers to purchase a new handset when they want to switch operators.
[9] See Klemperer (1989) for a model of sequential entry in the presence of switching costs. Van De Wielle and Verboven (2000) compare simultaneous and sequential entry in a model with switching costs.

One may thus expect stronger effects on mobile competition when competition is introduced sequentially than when competition is introduced simultaneously (though in the latter case the competition effects obviously take place at an earlier date). Second, when entry is sequential, switching costs may lead to pre-emptive behaviour. The incumbent firm may already start pricing or advertising aggressively in the period prior to entry.

4.2.3 Diffusion models

This section provides a basic description of the modelling approach used in this chapter. Technological innovations, such as mobile telecommunications, are typically not immediately adopted by all potential consumers. Consumers are differentiated in their preferred timing of adoption, so that a gradual diffusion of innovations may be expected. Various alternative diffusion models have been used to describe this process. Out of these, the 'epidemic' approach proved to be particularly popular, as it fitted the diffusion path of many innovations remarkably well. The adoption of innovation by the different agents is modelled in a similar way as diseases spread in biology: in other words, the flow of the adopters of the new technology is related to the stock of existing adopters. This relationship can be many-fold. The most common class of diffusion functions, in particular the logistic diffusion function, are S-shaped or sigmoid.

Econometric analysis of the diffusion of innovation first looked at the agricultural sector. Griliches (1957) pioneered an 'epidemic' diffusion model to study the diffusion of hybrid corn of the following type:[10]

$$y_t = y^*/[1 + e^{-(a+bt)}] \tag{4.1}$$

where y_t is the number of the agents that have adopted the new technology at time t. The number of total potential adopters is y^*.

Taking the first derivative, the flow of new adopters at time t is

$$\frac{dy_t}{dt} = by_t(1 - \frac{y_t}{y^*}) \tag{4.2}$$

The parameter b indicates the *diffusion speed*, which is constant in the case of the logistic diffusion function. Sigmoid diffusion functions have the characteristic of the second derivative being positive first and negative

---

[10] Chow (1967) applied such a model to study the diffusion of computers. An alternative model has been proposed by Bass (1969). For surveys, see Davies (1979), Stoneman (1983) and Geroski (2000).

thereafter, with an inflection point in between. In the case of the logistic diffusion function, the inflection point is at $y^*/2$. In other words, the growth rate of adopters is the largest when half of the population has adopted.[11]

In this model, as in other diffusion models, the flow of new adopters of the technology is related to the stock of existing adopters. When the stock of existing adopters is small, there is little risk of 'contagion'. As the stock increases, the risk of contagion increases, implying an exponential rise in the flow of new adopters. As the stock comes closer to the total number of potential adopters, the flow of new adopters gradually decreases and eventually becomes zero. The diffusion of the new technology thus follows an S-shaped function.

Sigmoid diffusion models, however, have the drawback that they are not able to give an economic explanation of the spread of new technologies. Diffusion is considered as an exogenously given process where choices by individual adopters are not modelled. However, there are also alternative approaches that take account of strategic interactions between firms in the adoption of new technologies.

These constant parameters are now combined with other variables that let us introduce more flexibility into the diffusion path. There are different approaches. Bewley and Fiebig (1988) proposed the parameters b as a function of a set of variables z_i as follows:

$$b_t(\mathbf{z}) = b_0 + \sum_i b_i z_{it} \tag{4.3}$$

Substituting into a logistic function gives:

$$\frac{dy_t}{dt} = b_t(\mathbf{z}) \, y_t (1 - \frac{y_t}{y^*}) \tag{4.4}$$

Through further rearrangements, this can be written as:

$$\frac{dy_t}{y_t \, dt} = b_0 - \frac{b_0}{y^*} y_t + \sum b_i z_{it} - \sum \frac{b_i}{y^*} z_{it} \, y_t \tag{4.5}$$

This model nests the static model as expressed by the first two terms on the right-hand side. The adaptive component depends on the exogenous

---

[11] There is another frequently used sigmoid diffusion function, the Gompertz diffusion curve. It is described by the function $y_t = y^* \exp\{-\exp\{-(a+b)\}\}$. Taking the first derivative, the flow of new adopters at time t is $\frac{dy_t}{dt} = b y_t$ (log y^* − log y_t). The inflection point occurs at $1/\exp\{1\}$ which is approximately at 37 per cent of the population of adopters. This diffusion curve describes processes where the maximum growth of adopters takes place more at the beginning than in the case of the logistic function (Chow, 1967; Dixon, 1980).

factors z_i. Moreover, the impact of the exogenous factors is proportional to the growth rate. Any given change in the exogenous factor will thus have a greater influence when the system is in its rapid adjustment phase than it will at either the beginning or end of the diffusion process. Parameter a can be modelled in a similar way. Let us now proceed to the model utilised in the present chapter.

4.3 The econometric model

4.3.1 A logistic model of diffusion

Let y_{it} denote the number of agents that have adopted the new technology in country i at time t; let y_{it*} denote the total number of potential adopters. The fraction of the total number of potential adopters in country i that has adopted before time t is specified by the logistic distribution function:

$$\frac{y_{it}}{y_{it}^*} = \frac{1}{1 + \exp(-a_{it} - b_{it}t)} \qquad (4.6)$$

The variable a_{it} in (4.6) is a location or 'timing' variable. It shifts the diffusion function forwards or backwards, without affecting the shape of the function otherwise. For example, when a_{it} is very high, we may say that country i at time t is very 'advanced' in its adoption rate. The variable b_{it} is a measure of the diffusion growth. This can be verified from differentiating (4.6) with respect to t, and rearranging:

$$\frac{dy_{it}}{dt} \frac{1}{y_{it}} = b_{it} \frac{y_{it}^* - y_{it}}{y_{it}^*} \qquad (4.7)$$

This implies that b_{it} equals the growth rate in the number of adopters at time t, relative to the fraction of adopters that have not yet adopted at time t. Equivalently, this says that the number of new adopters at time t, relative to the fraction of adopters that have not yet adopted at time t, is a linear function of the total number of consumers that have already adopted at time t. This reflects the 'epidemic' character of the logistic diffusion model.

It can be verified that the second derivative of (4.6) is positive for $y_{it}/y_{it*} < 1/2$, and negative if the reverse holds. The diffusion of the number of adopters thus follows an S-shaped pattern, with a maximum diffusion speed reached when half of the total number of potential adopters has effectively adopted the new technology.

In the econometric analysis (4.6) is transformed as follows:

$$\log\left(\frac{y_{it}}{y_{it}^* - y_{it}}\right) \equiv z_{it} = a_{it} + b_{it}t \tag{4.8}$$

The dependent variable, z_{it}, is the logarithm of total number of adopters relative to the number of potential adopters that have not yet adopted. Equation (4.8) shows that this measure for the level of adoption evolves linearly through time. Three essential elements determine the diffusion of mobile telecommunication services: the total number of potential adopters, y_{it^*} (entering in z_{it}); the location variable, a_{it}; and the growth variable, b_{it}. We specify these elements in turn.

The total number of potential adopters, y_{it^*}

Assume that y_{it^*} evolves proportionally to the total population, POP_{it}. For example, one may specify:

$$y_{it^*} = \gamma_i POP_{it} \tag{4.9}$$

where γ_i is the proportion of the population in country i that will eventually adopt a mobile phone. In principle, the parameter γ_i can be estimated as having fixed effects for each country. In practice, it is difficult to estimate these fixed effects, since most countries are still at the early stages of diffusion. Gruber and Verboven (2001a), estimating the diffusion of mobile telecommunications in the EU, resolved this problem by pooling the data, and estimating a parameter γ, common for all countries. This facilitates estimation because one can exploit information from countries in both early and in more mature stages of diffusion. This approach may be justified in their study, which considered the relatively homogeneous group of EU countries. However, looking at a heterogeneous dataset covering almost all countries in the world, this approach is harder to justify. A more flexible approach would be to allow the parameter γ to differ across certain groups of countries, according to various economic and social determinants such as income, the level of education or urbanisation. In practice, this approach proves difficult, in part because within each group only a few countries had reached more mature stages of diffusion.[12]

An alternative approach is followed by Dekimpe, Parker and Sarvary (1998). Instead of estimating the total number of potential adopters, they treat it as a 'known' parameter. More specifically, based on industry interviews, they specify the total number of potential adopters as 'the percentage of the literate people living in urban areas having a sufficient

[12] The problems were of two types. First, convergence was often difficult to obtain, since the model is non-linear and the parameter γ often causes the term within the logarithm to become negative. Second, if convergence was reached, the standard errors were quite large, essentially suggesting the data were uninformative about the total market potential.

income to afford basic telephone service'. The present approach is in a similar spirit: it treats the fractions γ_i as known parameters, dependent on urbanisation and economic development. Since understanding the market potential is not the aim of this chapter, the main concern is to check the robustness of the findings with respect to alternative assumptions.

The location and growth variables, a_{it} and b_{it}

The location variable a_{it} and the growth variable b_{it} in (4.8) can be specified in a general form as follows:

$$a_{it} = \alpha_i^0 + \sum_{j=1}^{J} \alpha^j D_{it}^j + x_{it}\alpha \tag{4.10}$$

$$b_{it} = \beta_i^0 + \sum_{j=1}^{J} \beta^j D_{it}^j + x_{it}\beta \tag{4.11}$$

The parameters α_i^0 and β_i^0 are country-specific location and growth effects. The variables D_{it}^j are dummy variables to capture the effect of certain events j. More specifically, let T_i^j denote the time of a certain event j in country i – for example, the time at which a first GSM operator was introduced in country i. The dummy variable D_{it}^j then equals zero for $t < T_i^j$, and equals one for $t \geq T_i^j$. The parameters α^j and β^j measure the effect of event j on the timing and growth variables; they are assumed to be the same across countries. The vector x_{it} includes continuous variables affecting the location or growth variables, e.g. *per capita* income.

Specifications (4.10) and (4.11) allow an event j to have an effect on both the location and growth variable in an unrestricted way. Most of the empirical literature implicitly imposes a structure on the specification by allowing the variable to enter only in the location or in the speed variable. The present approach proposes imposing some more systematic structure by assuming that there is no discontinuous change in the number of adopters after event j takes place. Event j may thus smoothly accelerate or decelerate the diffusion of innovation. More formally, the adoption level at the time of introduction of event j (i.e. at T_i^j) is equal to the adoption level slightly before the time of introduction of event j (i.e. at $T_i^j - \varepsilon$ (with ε small)). Since at T_i^j, $D_{it}^j = 1$, and at $T_i^j - \varepsilon$, $D_{it}^j = 0$, this condition implies that:

$$\alpha_i^0 + \alpha^j + \sum_{k \neq j} \alpha^k D_{it}^k + x_{it}\alpha + \left(\beta_i^0 + \beta^j + \sum_{k \neq j} \beta^k D_{it}^k + x_{it}\beta \right) T_i^j$$

$$= \alpha_i^0 + \sum_{k \neq j} \alpha^k D_{it}^k + x_{it}\alpha + \left(\beta_i^0 + \sum_{k \neq j} \beta^k D_{it}^k + x_{it}\beta \right) T_i^j \tag{4.12}$$

which simplifies to:

$$\alpha^j = -\beta^j T_i^j \tag{4.13}$$

Substituting (4.11), using restriction (4.13), into the transformed diffusion equation (4.8), the following obtains:

$$z_{it} = \alpha_i^0 + x_{it}\alpha + \left(\beta_i^0 + x_{it}\beta\right)t + \sum_{j=1}^{J} \beta^j D_{it}{}^j\left(t - T_i^j\right) \tag{4.14}$$

The data fails to reject the restricted equation (4.14) against the more general equation without restriction imposed (4.13). The focus of the attention is thus on (4.14) as the econometric reference model of the diffusion process.

4.3.2 Econometric specification

The following discusses how to include the variables referring to technology and competition in the econometric model of diffusion. First, it is explained how the role of the timing of first-entry licences is treated. Then the effects of competition, technological systems, competition between technological systems and country characteristics are explained in more detail.

The timing of first-entry licences

Equation (4.14) was first estimated without imposing any structure on the country-specific location and growth fixed effects α_i^0 and β_i^0. An interesting hypothesis is whether there is a relationship between these country-specific effects. For example, is it possible that an 'advanced' country (high-location effect) experiences a lower growth rate than a country that is lagging behind (low-location effect)? To the extent that this is the case, there is catching-up by latecomers, or international convergence. This may occur for several reasons, such as declining investment costs through calendar time, international learning spillovers, etc. One simple way to incorporate a catching-up effect is by imposing the following relationship between α_i^0 and β_i^0:

$$\beta_i^0 = \beta^0 - \lambda\alpha_i^0 \tag{4.15}$$

If late-coming countries catch up, then the parameter λ is positive. Substituting this expression into (4.14), it can be verified that all countries converge to the same number of adopters (holding all other variables constant) at time $t = 1/\lambda$. Hence, the inverse of the parameter λ may be interpreted as the time at which countries converge. This specification for a catching-up effect was imposed in the study by Gruber and Verboven (2001a). Note that this restriction may be overly restrictive in a study covering a more heterogeneous group of countries: if countries converge at all, they presumably do not converge at the same time.

There is a more flexible way to take into account the fact that late-coming countries may catch-up, to a partial extent, with early countries. In particular, consider a specification in which countries converge at time $t = 1/\lambda$ for only a *fraction* σ of the difference in initial adoption levels. At the time of introduction, $t = T_i^0$, country i realises an initial adoption level $z_{iT_i^0} = \alpha_i^0 + \beta_i^0 T_i^0$. A generalisation of (4.15) is:

$$\beta_i^0 = \beta^0 - \lambda\left(\alpha_i^0 - \sigma\left(\alpha_i^0 + \beta_i^0 T_i^0\right)\right) \tag{4.16}$$

Substituting this expression into (4.14), it can be verified that at $t = 1/\lambda$ countries converge to the same adoption level up to a fraction σ of the initial diffusion level $\alpha_i^0 + \beta_i^0 T_i^0$. For example, if $\sigma = 0$, then convergence is as in (4.15); in contrast, if $\sigma = 1$, then there is convergence at $t = 1/\lambda$, except for any possible differences in adoption levels at the time of introduction. Note that (4.16) still implies, independent of σ, that two countries i and j, which start adopting at a different time, converge to the same adoption level at $t = 1/\lambda$, provided they are equally 'advanced' (i.e. $\alpha_i^0 = \alpha_j^0$) – i.e. have the same initial adoption level.

To clarify this, figure 4.1 plots the (transformed) diffusion curve for three countries when (4.16) holds and $\sigma = 1$. Country 1 and 2 start at a different introduction date, but at the same level. They fully converge at $t = 1/\lambda$. Country 1 and 3 have a different adoption date and also start at a different level. They converge at time $t = 1/\lambda$, up to the initially different level. Countries 2 and 3 start at the same date but at a different level; they do not converge.

The timing of additional-entry licences

The effects of additional entry licences can be taken into account through several dummy variables D_{it}^j. A distinction is made between introducing competition among analogue operators and introducing competition between digital operators. Furthermore, one can distinguish between simultaneous entry, where two or more operators enter at once, and sequential entry, where one operator enjoys a monopoly period before additional entrants enter. Finally, a distinction is made between an initial effect of competition on diffusion growth and the effect after one year. The reasons for including all these variables have already been discussed. Let us now define the various dummy variables more precisely.

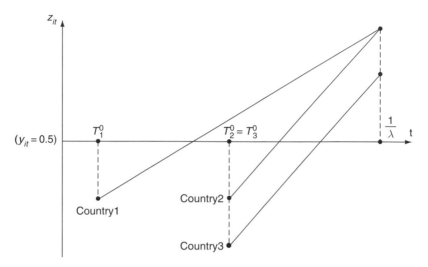

Figure 4.1 *The transformed diffusion curve*
Note: This figure shows the diffusion for three countries under (4.16), assuming that $\sigma = 1$.
Country 1 and 2 start at a different introduction date, but at the same diffusion level. They
fully converge at $t = 1/\lambda$. Country 1 and 3 start at a different adoption date and also at a
different diffusion level. They converge at $t = 1/\lambda$, except for the full amount of the initially
different level (since $\sigma = 1$). Country 2 and 3 start at the same date but at a different level. They
do not converge (since $\sigma = \lambda$). To depict situations where $\sigma < 1$, the curve for country 3 needs
to be modified: the end point increases until it reaches the end point of the other two countries
for $\sigma = 0$.

- *COMP_A, COMP_D*: dummy variables equals 1 as soon as competition
  between analogue or digital operators is introduced
- *SIMCOMP_A, SIMCOMP_D*: dummy variables equals 1 as soon as
  simultaneous competition between analogue or digital operators is
  introduced
- *SEQCOMP_A, SEQCOMP_D*: dummy variable equals 1 as soon as
  sequential competition between analogue or digital operators is introduced.
  In addition to these dummy variables there are also one-period leads,
  i.e. *SIMCOMP_A(+1), SIMCOMP_D(+1), SEQCOMP_A(+1)* and
  SEQCOMP_D(+1). When these lead variables are included, the original
  four variables measure the effect on the growth of adoption during the first
  year of (analogue/digital; simultaneous/sequential) competition; the lead
  variables measure the effect on the annual growth from the second year
  onwards. In other words, these variables measure whether the effects of
  competition occurred mainly in the first year, or also persisted in later
  years. For the sequential competition variables, there are also one-period
  lags: *SEQCOMP_A(−1)* and *SEQCOMP_D(−1)*. These lagged variables

measure the effect on the growth of adoption in the year prior to competition. The lags and leads can tell us something about whether competition had temporary or enduring effects.

Technological systems and competition between them

For each country, the effect of the switch from analogue to digital technology can be summarised through the following variable:

- *DIGITAL*: dummy variable equals 1 if a digital system has been introduced.

The effect of diffusion when different technological systems compete with each other is captured by the following dummy variables:

- *COMPSYST_A, COMPSYST_D*: dummy variable equals 1, if there are two or more competing analogue or digital systems. Since the variable DIGITAL is included, COMPSYST_D measures the additional effect of competing digital systems relative to the independent effect of the digital technology.

To capture the effects of diffusion when a digital system is introduced without a previously introduced and coexisting analogue system, the following variable has been defined:

- *SINGLE_D*: dummy variable equals 1 if a digital system is introduced without a previously introduced and coexisting analogue system.

Country characteristics

The following variables are included in the vector x_{it}, referring to country characteristics affecting the timing and speed of innovation:

- *GDPCAP*: income per head, measured as real GDP *per capita* converted into US dollars. This variable is expected to have a positive impact on the diffusion of innovation.
- *MAINCAP*: the number of fixed mainlines *per capita*. This variable captures the size of the fixed network and may have a positive or a negative effect, depending on whether adopters view mobile telecommunications services as a complement or a substitute for a fixed connection.
- *WAITLIST*: the ratio between registered applications for a fixed line and the number of connected fixed line subscribers. This variable thus measures the waiting list for a fixed line connection and captures the level of efficiency of the fixed operator, as well as the current 'excess demand' for telecommunication services. It is expected to have a positive impact on the diffusion.

4.3.3 Data description

The econometric estimates are based on annual data and cover 140 countries that have adopted cellular telecommunications. The data set covers the entire evolution of the cellular mobile industry (1981–97)[13] for most countries in the world. Apart from the countries that have not adopted cellular telecommunications, this sample excludes twenty-two adopters which are mostly very small countries. In total, the sample represents 94 per cent of the world's population. The time series starts in 1981 and therefore covers all cellular markets from the first year, with the exception of Japan where mobile telecommunications were introduced in 1979. The data on the number of analogue and digital subscribers, the waiting list and the number of fixed mainlines are from the World Telecommunications Indicators of the ITC, (1999). The information about the type of system is gathered from various sources, such as the trade press (*Mobile Communications* and EMC), GSM MoU (http://www.gsmworld.com), Bekkers and Smits (1997) and Garrard (1998). Macroeconomic data such as GDP and population are taken from the World Bank's *World Development Indicators*. Table 4.1 presents some descriptive statistics on the diffusion levels at different points in time.

4.4 Empirical results

After adding an error term, the diffusion model (4.14) was estimated using (non-linear) least squares. Table 4.2 lists the results. Estimating (4.14) produced a good fit of the data ($R^2 = 0.98$), but the standard errors of most coefficients were relatively high. However, adding restriction (4.13), which imposes a continuous change after a new event, for any specification, improved the significance of parameters. This restriction could not be rejected by the data. The reason for this can be shown by plotting the growth effects against the location effects, as drawn in figure 4.2. This figure indicates a strong negative relationship: advanced countries (with a high location effect) have a strong tendency to grow more slowly than countries that are lagging behind (with a low-location effect). This negative relationship was the reason for estimating the model under restrictions (4.15) or (4.16), testing for the presence of a catching-up effect, or international convergence.

Column (i) in table 4.2 shows estimates when country characteristics are excluded and constraint (4.15) is applied (i.e. $\sigma = 0$ (full international convergence at estimated time $t = 1/\lambda$)). Column (ii) shows estimates when country characteristics, *GDPCAP, MAINCAP* and *WAITLIST*, are also included. Column (iii) allows λ to vary across groups of countries. Column (iv) generalises (4.15) to the more flexible constraint (4.16).

---

[13] The exception is Japan, where cellular mobile telecommunications had already been introduced in 1979, see below.

Table 4.1 *Descriptive statistics: mobile penetration rates*[a]

	Number of observations	Average	St dev.	Min.	Max.
		After first full year of introduction			
All countries	139	0.3	0.6	0.0	6.5
LDC[b]	75	0.1	0.3	0.0	1.8
MDC[b]	64	0.4	0.9	0.0	6.5
		After fifth full year of introduction			
All countries	91	1.7	2.3	0.0	12.6
LDC	31	0.9	1.7	0.0	8.2
MDC	60	2.1	2.5	0.1	12.6
		After tenth full year of introduction			
All countries	36	6.2	5.5	0.0	26.4
LDC	6	0.7	1.1	0.0	2.8
MDC	30	7.3	5.4	0.4	26.4
		End of 1985			
All countries	140	0.0	0.2	0.0	1.5
LDC	76	0.0	0.0	0.0	0.0
MDC	64	0.1	0.3	0.0	1.5
		End of 1990			
All countries	140	0.4	0.9	0.0	5.4
LDC	76	0.0	0.2	0.0	1.8
MDC	64	0.8	1.2	0.0	5.4
		End of 1995			
All countries	140	2.6	4.5	0.0	22.7
LDC	76	0.4	1.4	0.0	10.0
MDC	64	5.2	5.4	0.1	22.7

Notes: [a] The penetration rate is calculated as number of mobile telecommunications subscribers per 100 inhabitants.
[b] LDC = Less developed countries (income class 1 and 2 according to World Bank Classification): MDC = More developed countries (income class 3 and 4 according to World Bank Classification).

The effect of country characteristics

Specifications (ii)–(iv) show that countries with a high income *per capita* (*GDPCAP*) tend to be more advanced in adopting mobile phones, yet the effect is diminishing over time. The overall effect of income on mobile penetration remains positive roughly until 2010 (for specification (ii)). This

Table 4.2 *Empirical results for diffusion equation (4.14)*

	(i)		(ii)		(iii)		(iv)	
λ	.029**	(0.001)	.027**	(.002)	.030**	(.002)	.046**	(.002)
λ_2					.000	(.002)		
λ_3					.006**	(.001)		
λ_4					.009**	(.002)		
σ							.776**	(.074)
β_0	.176**	(.017)	.194**	(.033)	.110**	(.026)	.233**	(.030)
Growth parameters for competition variables								
COMP_A	.059**	(.017)	.039*	(.020)	.032*	(.017)	.037*	(.021)
COMP_D	.155**	(.041)	.134**	(.045)	.119**	(.043)	.181**	(.045)
Growth parameters for technology variables								
DIGITAL	.059	(.039)	.067*	(.044)	.086*	(.043)	−.005	(.044)
SINGLE_D	.055	(.078)	.061	(.085)	.013	(.078)	−.110	(.092)
COMPSYST_A	−.036	(.024)	−.044*	(.026)	−.046*	(.023)	.002	(.027)
COMPSYST_D	−.198	(.151)	−.131	(.189)	−.009	(.184)	−.182	(.184)
Location parameters for country characteristics								
GDPCAP			.143**	(.027)	.142**	(.031)	.101**	(.022)
MAINCAP			.747**	(.260)	1.091*	(.347)	.925**	(.197)
WAITLIST			−.237	(.222)	−.090	(.236)	−.413*	(.198)
Growth parameters for country characteristics								
GDPCAP			−.005**	(.001)	−.007**	(.002)	−.001	(.001)
MAINCAP			−.029**	(.012)	−.053**	(.018)	−.026**	(.008)
WAITLIST			.052**	(.013)	.048**	(.014)	.066**	(.014)

Notes: * Statistically significant at 5 per cent level.
** Statistically significant at 1 per cent level.
Standard errors in brackets

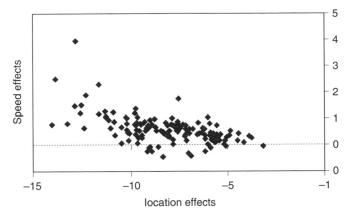

Figure 4.2 *Estimated country-specific fixed effects*

is intuitive, given the large fraction of the budget spent on a mobile phone during the early years, and the declining prices afterwards. Similarly, countries with a large fixed network (*MAINCAP*) tend to be more advanced in adopting mobile phones. Yet again the effect is diminishing over time and becomes negligible around 2007 (for specification (ii)). This suggests that the fixed network is largely viewed as a complement to mobile phones. Finally, countries with a large waiting list for a fixed line connection initially have lower mobile penetration levels. Yet these countries experience a very strong and significantly higher annual growth rate than countries with a low waiting list. This brings them to more advanced adoption levels from 1987 onwards. Mobile telecommunications may thus be a very suitable tool for providing telecommunications access in inefficient fixed line markets – i.e. more typically developing countries.

The timing of first-entry licences

Consider now the relevance of the timing of first-entry licences, by looking at how fast early and late-coming countries converge. Columns (i) and (ii), which impose restriction (4.15), find a highly significant estimate of λ of 0.029 and 0.027, respectively. Countries that are less advanced in the level of adoption thus catch-up by growing faster than early countries. Nevertheless the catching-up effect is very slow: the date of convergence in adoption levels ($t = 1/\lambda$) is $t = 34.5$ and $t = 37.0$, in the specifications under column (i) and (ii), respectively (with standard errors of 2.1 and 5.1, respectively). Because $t = 0$ corresponds to 1980, this means that countries would converge in 2014 and 2019, respectively (with 95 per cent confidence intervals of 2010–18 and 2009–29, respectively). The estimated convergence dates are later than in Gruber and Verboven (2001a), who found convergence within the countries of the EU around 2008. This is not surprising given that the set of countries in the present sample is much more heterogeneous than the EU countries.

To incorporate this heterogeneity between countries, restriction (4.15) has been relaxed in two ways. First, λ was allowed to vary across the following four groups of level of economic development indicated by income *per capita* (according to the World Bank classification): low, lower-middle, upper-middle and high-income countries. The estimates in column (iii) show that there are indeed significant differences in catching-up across the four groups of countries. The least developed group 1 and group 2 countries show the slowest convergence (around 2013), preceded by group 3 countries (around 2008) and group 4 (around 2006). Even though late-coming countries thus catch-up faster if they come from more developed countries (group 3 and group 4), the delay is still substantial.

The second way of relaxing restriction (4.15) was by allowing for partial convergence, using (4.16) instead of (4.15). In other words, convergence may occur except for a fraction σ of the difference in initial adoption levels. This allows us to focus on convergence between countries with different introduction dates that are otherwise similar (see figure 4.1). The results in column (iv) now show an estimate of λ equal to 0.046 and an estimate of σ equal to 0.78. According to this specification, countries converge around 2002, except for 78 per cent of the possible difference in the initial adoption level. Referring to figure 4.1, this means that two countries issuing a first licence at a different point in time, but with the same initial level, converge around 2002, whereas countries with different initial adoption levels show little convergence. Intuitively, the effects of regulatory delay in issuing first licences persist until 2002.[14] Any remaining lack of convergence after that time follows from persisting initial differences in adoption levels across countries.

The introduction and timing of additional-entry licences

All the specifications in table 4.2 consider the effects of introducing two or more competing operators during the analogue and during the digital era. One can see that introducing competition between operators had a significant impact on the growth of mobile diffusion. The effect was especially significant during the digital era, and less pronounced during the analogue era. This is consistent with the hypothesis that *capacity* plays a major role in explaining the magnitude of the competition effects. During the analogue era, capacity was constrained, thereby mitigating the positive effects from competition. Empirical evidence suggests that prices remained relatively high indeed after the introduction of competition during the analogue era.

[14] Some studies indicate high welfare costs from regulatory delays in issuing licences, e.g. Hausman's (1997) estimate that regulatory delay in analogue cellular licences cost the US economy up to $100 billion.

As discussed earlier, consumer switching costs may have an additional influence on the effects from competition. To investigate this, table 4.3 extends specification (i) of table 4.2 to explore the competition effects in further detail, comparing simultaneous with sequential entry and distinguishing between pre-emptive, immediate and future competition effects. Column (i) compares the effects of simultaneous and sequential entry ($SIMCOMP\_A$ vs. $SEQCOMP\_A$, and $SIMCOMP\_D$ vs. $SEQCOMP\_D$). It can be seen that the impact on the diffusion of mobile adoption was substantially stronger when entry was introduced sequentially than when it was introduced simultaneously. The sequential entry effect is especially strong during the digital era when capacity is larger, but it is also present during the more capacity-constrained analogue era. One explanation for the stronger sequential entry effect is that the mobile market is still growing: since competition is on average introduced at a later date under sequential entry, some catching-up may be expected. An alternative explanation is that a new sequential entrant needs to price rather aggressively to obtain at least some market share if the incumbent's consumers face significant switching costs.

To further explore the role of switching costs, we considered pre-emptive, immediate and future competition effects. Column (ii) of table 4.3 distinguishes between the competition effect during the first full year of competition and the effect afterwards, by introducing a lead variable of the competition variable. Interestingly, it can be seen that most of the competition effect takes place during the first year. For simultaneous entry during the analogue and the digital eras, the competition effects are 0.793 and 0.271, respectively, during the first year of competition, and drop to insignificant numbers of $0.793-0.855 = -0.062$ and $0.271-0.222 = 0.049$, respectively, afterwards. For sequential entry during the analogue period, the competition effect is 0.713 during the first year of competition, and drops to an insignificant number of $0.713-0.703 = 0.010$ afterwards. Only for sequential entry during the digital era does the competition effect remain large after the first year $(0.631-0.026 = 0.605)$. Yet this is because for this particular case our sample has few years of observations after the second entrant has entered.

The fact that competition mainly influences diffusion during the first year is consistent with our hypothesis that consumers have *switching costs*. During the first year, firms compete vigorously to build up market share to exploit market power in future stages. Once an installed base is built up, competition becomes 'softer'. Note that competition does not become so 'soft' as actually to lower the adoption level or reduce the adoption growth below the pre-competition rate. This is because in this market there new consumers appear to compete for in every period.

Table 4.3 Simultaneous vs. sequential entry effects and technological systems' competition

	(i)		(ii)		(iii)		(iv)	
Analogue technology								
SIMCOMP_A	.018	(.023)	.793**	(.267)	.800**	(.266)	.800**	(.266)
SIMCOMP_A(+1)			−.855**	(.293)	−.860**	(.292)	−.859**	(.292)
SEQCOMP_A(−1)					.297*	(.166)	.394**	(.065)
SEQCOMP_A	.112**	(.023)	.713**	(.122)	.197	(.309)		
SEQCOMP_A(+1)			−.703**	(.141)	−.479**	(.183)	−.377**	(.084)
COMPSYST_A	−.073**	(.025)	−.060**	(.025)	−.065**	(.026)	−.065**	(.025)
Digital technology								
SIMCOMP_D	.141**	(.041)	.271**	(.066)	.299**	(.067)	.298**	(.066)
SIMCOMP_D(+1)			−.222*	(.106)	−.231*	(.105)	−.231*	(.105)
SEQCOMP_D(−1)					.447**	(.188)	.411**	(.098)
SEQCOMP_D	.465**	(.096)	.631**	(.185)	−.082	(.353)		
SEQCOMP_D(+1)			−.026	(.024)	−.008	(.025)	−.013	(.018)
COMPSYST_D	−.129	(.151)	−.083	(.149)	−.088	(.148)	−.091	(.148)

Notes: * Statistically significant at 5 per cent level.
** Statistically significant at 1 per cent level.
Standard errors in brackets

Columns (iii) and (iv) of table 4.3 investigate whether, in the case of sequential entry, incumbents have an incentive to pre-empt in the period prior to actual entry. This may be done, for example, through limit pricing (charging lower prices than a monopolist) or by following aggressive marketing campaigns.[15] While switching costs may explain the incumbent's limit pricing as a strategy to build up a market share to exploit future market power, it is not the only possibility. In fact, an incumbent may 'limit overprice' (charge higher prices than a monopolist) if it is more important to induce 'soft' competition by the future entrant. Limit pricing is more likely if switching costs are present but not too large, and if there is a significant growth of new consumers.

To assess the presence of pre-emptive behaviour, a lagged dummy variable for the sequential entry variables was included. Column (iii) shows that this lagged variable has a significant and stronger effect during both the analogue and the digital eras. This suggests the presence of pre-emptive behaviour by incumbent firms, through limit pricing, aggressive marketing campaigns or otherwise. To obtain further insights, column (iv) constrains the effect of the lagged (pre-emption) competition variable to be the same as the actual competition variable. This shows more precisely how the diffusion level in the analogue era increased, especially during the year preceding competition and the year of actual competition.[16]

To summarise this analysis, these estimates suggest that competition has a stronger impact during the digital era than during the analogue era, thanks to drastically increased capacity. Moreover, competition induces diffusion especially during the early years (or even in the preceding year in the case of sequential entry), and sequential entry has a stronger impact than simultaneous entry. This is consistent with the presence of consumer switching costs, accounting for the fact that new consumers appear in every period.

The role of technological systems and systems competition
Tables 4.2 and 4.3 also include an assessment of the effects of different technological systems. First, note that the presence of a digital

---

[15] An example in this respect is the UK, discussed in chapter 3. In spite of a duopoly during the second half of the 1980s and the early years of the 1990s, prices for mobile telecommunications stayed constant in nominal terms. Only the sequential entry of two further firms in 1993 and 1994, respectively, induced a pattern of falling prices (Valletti and Cave, 1998).

[16] The fact that for analogue sequential entry the competition effects are lower in specification (iii) and (iv) as compared to specification (ii) does not mean that the results are not robust. This is because (ii) does not take into account pre-emptive effects. Properly to compare (ii) with (iii) and (iv), one should add the pre-emptive and actual effects in specifications (iii) and (iv). One then obtains a similar cumulative effect.

technology (*DIGITAL*) has only a modest independent impact on diffusion growth. Quite intuitively, the beneficial capacity impact of the digital technology works best in those cases where it has been combined with the introduction of competition (*COMP_D*), as discussed earlier. Similarly, the introduction of the digital technology without a preceding analogue period (*SINGLE_D*) had no significant independent impact. This suggests the absence of a lock-in effect in the less efficient analogue system.

Now consider the effects on the diffusion growth when there were two or more competing analogue or digital systems, measured by *COMPSYST_A* and *COMPSYST_D*. Table 4.3 suggests that competition between analogue systems (e.g. NMT and TACS) slowed down the growth in mobile diffusion. This is confirmed by large and significant negative annual growth effects of about 6–7 per cent in the more elaborate specifications of table 4.4.[17] Competition between digital systems (GSM and non-GSM) also seemed to slow down diffusion. While the negative point estimates for the effect of digital systems' competition seem quite substantial, they are also rather imprecise. This is because there are only a few observed cases.

Table 4.4 reports the estimates of the basic equations as in table 4.2, but with the inclusion of dummy variables for the technologies NMT, TACS, AMPS, C-450 and GSM. The estimates show that some technological systems significantly affect diffusion growth, relative to this benchmark. The analogue NMT and especially TACS technological standards significantly slow down the growth of diffusion. The digital non-GSM standards significantly accelerate the growth of diffusion in most specifications. The effect of the digital GSM standard is only modest and usually insignificant. This latter finding contrasts with common wisdom and policy reports on the successes of the GSM standard. With this specification, the effect of competition between analogue systems on slowing down the growth in mobile diffusion is even stronger and there are significant negative effects (between –5 and –14 per cent). Again, the effect of competition among digital systems is not significant.

To interpret these results on competition among systems, recall that there may be both advantages and disadvantages from having competing systems rather than single standards. The major advantage of allowing competing systems is that markets may not be locked in to inferior technologies and that firms are motivated to continuously invest in R&D to improve the quality of their technology. The major disadvantages of

---

[17] Even stronger and significant negative effects (between –5 and –14 per cent) were obtained in specifications that distinguished between the different quality effects of the NMT, TACS, AMPS and C-450 analogue technologies. To simplify the exposition, we do not report the results of these specifications.

Table 4.4 Empirical results for diffusion equation (4.6), including technology dummies

	(i)	(ii)	(iii)	(iv)
λ	.027** (.001)	.024** (.003)	.028** (.003)	.043** (.003)
λ_2			.0004** (.0016)	
λ_3			.0051** (.0016)	
λ_4			.0086** (.0018)	
σ				.839 (.090)
β_0	.277 (.031)	.283 (.058)	.186 (.051)	.307 (.050)
Growth parameters for competition variables				
COMP_A	.050** (.020)	.033 (.023)	.033 (.021)	.055** (.025)
COMP_D	.166*** (.043)	.152** (.047)	.136** (.044)	.162** (.046)
Growth parameters for technology variables				
NMT	−.090** (.027)	−.068** (.030)	−.044 (.028)	−.024 (.028)
TACS	−.142** (.030)	−.130** (.036)	−.084** (.034)	−.090** (.035)
AMPS	−.070** (.032)	−.012 (.036)	−.007 (.035)	−.034 (.037)
C-450	−.018 (.044)	.031 (.050)	−.012 (.046)	.081 (.053)
GSM	.015 (.038)	.021 (.042)	.056 (.042)	−.049 (.042)
NONGSM	.137 (.118)	.396** (.140)	.443** (.136)	.482** (.137)
COMPSYS_A	−.134** (.038)	−.106** (.041)	−.108** (.042)	−.054 (.039)
COMPSYS_D	−.090 (.157)	.131 (.197)	.284 (.197)	−.055 (.194)
SINGLE_D	.015 (.097)	.094 (.104)	.110 (.100)	−.100 (.110)

Table 4.4 (cont.)

	(i)	(ii)	(iii)	(iv)
Location parameters for country characteristics				
GDPCAP		.107 (.027)	.116 (.030)	.071 (.022)
MAINCAP		1.130 (.258)	1.508 (.345)	1.195 (.197)
WAITLIST		−.276 (.212)	−.156 (.224)	−.481 (.191)
Growth parameters for country characteristics				
GDPCAP		−.004** (.001)	−.006** (.002)	−.006 (.009)
MAINCAP		−.005** (.001)	−.073** (.018)	−.036** (.008)
WAITLIST		.058** (.013)	.056** (.014)	.072** (.014)

Notes: * Statistically significant at 5 per cent level.
** Statistically significant at 1 per cent level.

allowing competing systems are that network externalities are more limited (especially when 'roaming' is valued highly) and that economies of scale in the manufacture of equipment are not fully exploited. Our empirical results thus indicate that the disadvantages of competing systems (network effects and scale economies) were dominant during the analogue era. During the digital era, the disadvantages may have been partly balanced by the advantages from technological systems' competition. This is consistent with the view of Shapiro and Varian (1999), who argue that the decentralised systems' competition approach followed in the USA may have hindered diffusion of the current technology, but gave the innovative CDMA technology a chance to develop: CDMA became the basis for 3G mobile telecommunication systems.

4.5 Conclusion

This chapter has looked at the effects of entry regulation and standard setting on the evolution of the cellular mobile telecommunications services industry, also controlling for a set of country-specific variables. It is shown that the policy design of market structure has to take account of technological constraints. One can distinguish between an analogue phase, during which the industry was potentially capacity-constrained, and a digital phase, during which these constraints were relaxed. Government policies affected the evolution of the industry in a different way during both phases.

First, the actual timing at which first-entry licences were issued had a significant impact on the diffusion of mobile services. The effects of regulatory delay in issuing first licences on cross-country differences in adoption levels were felt until around 2002. After that time, a lack of convergence was attributed to persisting initial differences in adoption levels.

Second, the introduction of second-entry licences (competition) also had a significant impact on the diffusion of mobile services. The effect was especially strong during the digital phase. This is consistent with the existence of binding capacity constraints during the analogue phase, compared to a drastically expanded capacity in the digital phase, confirming the expectation that competition speeds up diffusion.

Third, the timing at which second licences were introduced turned out to be very relevant. Simultaneous entry had a modest (but significant) impact on the diffusion, whereas sequential entry had a stronger impact, especially during the digital phase. Most of the competition effect took place during the first year of competition. In the case of sequential entry, the competition effect also took place in the year prior to second entry, indicating pre-emptive behaviour by the incumbent. These findings can be explained by strategic behaviour by the operators in the presence of consumer switching costs.

Finally, setting technology standards rather than allowing multiple competing systems was a relevant determinant of the evolution of the industry. Results suggest that a single analogue standard helped to develop the market significantly faster compared to competing analogue systems. This was consistent with the presence of network effects and scale economies. Imposing a single digital standard (e.g. GSM in the EU) also seemed to stimulate diffusion, yet the effect was imprecisely estimated; a longer time horizon might be required to assess whether the advantages from systems competition in the digital era (e.g. the emergence of the new CDMA system to be used for 3G mobile telecommunications) were outweighed by the network and scale advantages from a single standard.

With respect to country characteristics, the data suggest that income *per capita* and the size of the fixed network had a positive (but declining) effect on the level of diffusion. The length of the waiting list for the fixed network also has a positive effect on the level of diffusion, suggesting that mobile telecommunications were a suitable alternative in providing telecommunications access in inefficient fixed line markets. One of the broader policy conclusions that can be drawn from this chapter is that *public policy decisions* typically have a persistent effect on the evolution of regulated industries. The cost of regulatory failure can therefore be very high. Firm entry and timing are important determinants for market evolution, and this importance increases as capacity constraints are relaxed.

5 Market conduct and pricing issues in mobile markets

5.1 Introduction

The mobile telecommunications industry is an interesting laboratory in which to study the market behaviour of firms with market power. This is the realm of *oligopoly theory*, which has made huge progress since the 1980s,[1] in particular thanks to the game theoretic approach. The matching of theory with empirical evidence has brought many new insights for the industrial organisation literature and has been proved very useful in providing normative purposes for regulators, in particular in designing markets. Firms have scope for reducing price competition by means of appropriate choice of product differentiation. Oligopoly theory proved to be a difficult subject because the equilibrium outcomes in oligopoly models are very sensitive to the behavioural assumption of firms. The problem is typically represented as a three-stage game in which firms (1) first decide on entry, (2) then on product characteristics and (3) finally on the nature of price competition. This chapter is not concerned with the entry decision (1), which will be considered in chapters 6 and 7, but it deals with stages (2) and (3), which are concerned with product specification and pricing behaviour, respectively.

The mobile telecommunications industry is a network industry where different networks compete for customers. Pricing decisions thus depend on *interaction with other networks* and are also constrained by *regulatory decisions*. The theoretical and practical foundations of pricing in telecommunications have changed significantly along the technological trajectory of the industry and its market structure. With the rolling back of monopolies, pricing became strongly influenced by the compensation mechanisms set up for interconnected traffic among networks. In contrast with many other network industries, the mobile telecommunications industry is less subject

---

[1] For an introduction to the theory of oligopoly, see Shaked and Sutton (1983) and Tirole (1988).

to specific price regulations, reflecting the idea that facility-based network competition is sufficient to restrain retail prices. Nevertheless, several market distortions (such as high termination and 'roaming' prices, as well as the emergence of firms with strong market power) could be observed. Regulatory challenges of redressing market failures related to the abuse of market power thus appear far from being resolved.

This chapter is organised as follows. Section 5.2 provides a survey of the theoretical issues in market conduct in oligopolies. Section 5.3 illustrates how product differentiation may conducted and section 5.4 surveys some empirical studies in this field that deal specifically with the mobile telecommunications literature. Section 5.5 deals with the theoretical principles of one-way and two-way access. These principles will be the benchmark for assessing the actual consequences of pricing regimes and regulatory action. In section 5.6, these principles will then be developed further and specifically for the context posed by the mobile telecommunications industry. This section will deal with how prices are set for the different services and where market power can most effectively be exploited. There are two separate issues in pricing: one is the question of who is paying for the call and the other is the determination of pricing. This leads to an exploration of the areas where regulatory action could be most effectively used (e.g. fixed to mobile access pricing and international 'roaming') and to what extent the use of different pricing principles – such as calling party pays (CPP) vs. receiving party pays (RPP) – may mitigate the abuse of market power. Section 5.7 describes the pricing policies adopted in submarkets of the mobile telecommunications industry, in particular in the wholesale and the retail sections of the market. Section 5.8 looks at the empirical evidence on the pricing trends observed in the industry and discusses whether these are in line with the principles set out previously. Section 5.9 concludes the chapter, discussing also the scope for regulatory action to remove market inefficiencies observed.

5.2 Theoretical considerations on market conduct

The recent theory of industrial organisation has devoted particular attention to the study of oligopolistic markets. Oligopoly theory tells us that product differentiation reduces the scope for substitution among products by consumers, and hence helps firms to relax the need for price competition. In this context, game theory has become the dominant framework for theoretical analysis. Oligopoly used to be a particularly difficult subject to study because the equilibrium outcomes in oligopoly models are very sensitive to the behavioural assumptions of firms. For instance, there is the well-known textbook case of arbitrarily small fixed entry costs leading to a monopoly outcome if firms compete on prices, but to an oligopoly outcome if firms compete on quantities (Cournot competition). It may prove useful to divide products into two broad categories: homogeneous products and differentiated products. The literature distinguishes three types of models of product differentiation: horizontal and vertical product differentiation and the goods-characteristics approach. Our focus will be on the former two types of product differentiation.[2]

5.2.1 Homogeneous products

Products are homogeneous when firms have little scope for attributing product characteristics to distinguish products supplied by different firms. The only distinguishing variable can be *price*. The consumer will buy the lowest-cost product. At equilibrium, products will be sold at the same price. With Bertrand competition, price is equal to marginal cost and profits are equal to zero. With Cournot competition, price is above marginal cost, with profits declining with the number of firms in the market.

The problem can be represented as a two-stage game. In the first stage, firms decide on market entry. In the second stage, they decide on price. Entry cost crucially affects market structure. With Bertrand competition, entry costs lead to a monopoly; with Cournot competition, market structure depends on the size of the entry cost. This can be represented as follows.

Consider the following inverse demand function, $p(Q) = s/Q$. s is the parameter for market size and Q are total quantities sold at market price, p. Assume constant marginal costs, c. It can be shown that in a Cournot equilibrium with n identical firms (where quantity supplied by each firm is $q = Q/n$) the equilibrium price is:

$$p = \frac{nc}{n-1} \tag{5.1}$$

Typically for a Cournot model, price is above marginal cost and declining with the number of firms. The fixed entry cost, F, sets an upper bound on entry. At Cournot equilibrium the profits for each firm are:

$$\Pi(n) = (p - c)q - F = \frac{s}{n^2} - F \tag{5.2}$$

---

[2] The goods-characteristics approach, pioneered by Lancaster (1966), defines goods as a 'bundle of characteristics' on which consumers express preferences. Unlike the two other types of models, consumers here may consume more than one good. Consumers are able to sum up characteristics, which may then be spread over a bundle of goods. This possibility of consuming a variety of goods makes this approach less appealing in cases where indivisibilities of consumption are important. For more recent developments, see Tirole (1988).

The Cournot equilibrium number of firms, n^*, is determined by the following zero profit entry condition:

$$\Pi(n^*, s, F) > 0 > \Pi(n^* + 1, s, F) \qquad (5.3)$$

For any $n > n^*$ industry, profits are zero. Fixed costs thus create an *entry barrier*. Neglecting the integer problem, one can derive the following expression:

$$n^* = \sqrt{\frac{s}{F}} \qquad (5.4)$$

One thus can derive relationships between the equilibrium number of firms, market size and entry costs: $dn^*/ds > 0$ and $dn^*/dF < 0$. The equilibrium number of firms thus increases with market size and decreases with fixed costs.

5.2.2 Horizontally differentiated products

The classical approach to horizontal product differentiation (Hotelling, 1929) considers competitive interaction as *product positioning*. Products may be identical, but they distinguish themselves by their position in the product space. Salop (1979) has proposed the following 'circular road' model. Assume that the product space has the form of a circle, and that firms position their products at an equal distance from each other on this circle. Once they have entered, firms compete on prices. Let z be the unit 'transport cost' for the consumer, n the number of firms and c the (constant) marginal cost. The symmetric equilibrium price is:

$$p = c + \frac{z}{n} \qquad (5.5)$$

and the profit for the individual firm is:

$$\Pi(n) = (p - c)q - F = \frac{sz}{n^2} - F \qquad (5.6)$$

As in the case of homogeneous products, market structure is also here a function of fixed costs. Again, neglecting the integer problem, one can derive the following expression for the equilibrium number of firms with zero profits:

$$n^* = \sqrt{\frac{zt}{F}} \qquad (5.7)$$

As in the case of the Cournot model with homogeneous products, market structure varies with market size and fixed costs. Moreover, increased transport costs lead to a larger number of firms. As firms have greater scope for raising prices, the market can support a larger number of firms at equilibrium. The central message of the horizontal product differentiation literature is that many equilibria can be obtained. These depend strongly on the entry order (Lane, 1980) and the presence of multiproduct firms.

5.2.3 Vertically differentiated products

With vertical product differentiation, *products differ in quality*. The distinctive feature is that all consumers prefer the same (higher)-quality product if the products are supplied at cost. Consumers are assumed to be distributed over a set of incomes, and the willingness to pay (WTP) for quality is determined by their income. The theoretical basis of vertical product differentiation models may be found in Gabszewicz and Thisse (1979, 1980), which was developed further by Shaked and Sutton (1982, 1983). Their models set up the following three-stage game. At stage one, firms decide on entry, at stage two they decide on which quality to supply and at stage three they compete *à la* Bertrand in prices. If the income distribution is limited to a certain range and if entry requires some small sunk cost $\varepsilon > 0$, then the only subgame perfect equilibrium in the three-stage game entails the entry of only two firms, which produce distinctive products and earn positive profits. Furthermore, the firm producing the higher-quality product enjoys higher profits than the firm supplying the lower-quality product. By moving away from each other through product differentiation, firms relax price competition, which otherwise would drive prices down to zero. Extending the range of income distribution increases the number of the firms that can be supported at equilibrium. Vertical product differentiation models may be used to explain certain forms of market structure such as a 'natural oligopoly'. Shaked and Sutton (1983) showed that there existed an upper bound, independent of market size, to the number of firms which coexist at equilibrium with positive market shares and prices exceeding unit variable cost. The authors refer to this as the Finiteness Property. The basic condition for this outcome is that unit variable cost must not increase too strongly with quality, so that the highest quality will still be preferred by all consumers if supplied at cost. The Finiteness Property is an interesting result; it prevents the industry from being fragmented as market size increases. This is not the case in the horizontal product differentiation model with sunk cost, where market shares are not bounded away from zero as market size increases. If instead Cournot competition is assumed at stage three of the game, then the Finiteness Property would not hold, since many firms would produce the same quality, and the number of firms would increase as market size increases. The result would be similar to the horizontal product differentiation case where we have unlimited entry when markets expand. Thus again, the equilibrium outcome crucially depends on the type of competition at stage three.

What has been said can be formulated in the following way. Let products be described by the characteristic *quality* denoted by the real number k where $k \in [a,b]$ with $b > a$. A higher k means higher quality and all customers agree on this if product is supplied at cost. Assume a continuum of customers with identical tastes but different incomes,[3] t. Income is distributed uniformly over the unit interval,[4] $t \in [0,1]$. Customers make indivisible and mutually exclusive purchases and buy at most one unit. The utility function is

$$U(t,k) = u(k) \cdot (t - p_k) \tag{5.8}$$

where $u(k)$ denotes the utility of consuming good of quality k with $u'(k) > 0$, and p_k denotes the price of product with quality k.

Consider the customer with income t_k such that $U(t_k, k) = U(t_k, k-1)$. This defines[5]

$$t_k = -\frac{u(k-1)}{u(k) - u(k-1)} p_{k-1} + \frac{u(k)}{u(k) - u(k-1)} p_k \tag{5.9}$$

as the income level of the customer indifferent between quality k at price p_k and quality $k - 1$ at price p_{k-1}. Customers with income higher than t_k strictly prefer quality k, and customers with income lower than t_k strictly prefer quality $k - 1$.

Define u_0 as the utility of not consuming the good, i.e.:

$$U(t,0) = u_0 \cdot t \tag{5.10}$$

Assume that each firm supplies one quality only. The above partitioning of income space also allows for a partitioning of firms in the quality space. Firms have the opportunity to choose different locations in the product space. Firms offering a high-quality product aim at rich customers, while firms offering an inferior-quality product aim at customers with lower income. Firm i thus has the following demand function:

$$x_k = (t_{k+1} - t_k) \cdot s$$

---

[3] Here, 'income' is to be understood as an indicator of the WTP.
[4] Shaked and Sutton (1982) assume a positive lower bound for the income distribution. Assuming the lower bound to be zero, as in the present case, eliminates a discontinuity in the first derivative of the profit function. In this way, the customer with zero income would never buy the good. Otherwise the consumer with the lowest income may strictly prefer to buy, or she may be indifferent.
[5] Strictly speaking, the customer should be indifferent between k and $k - \varepsilon$, where ε is any positive number indicating the next closest lower quality available. I will use the notation $k - 1$ to ease the overall notation.

With cost equal to zero, firm i supplying quality k enjoys revenue

$$R_{ik} = p_i \cdot (t_{k+1} - t_k) \cdot s \tag{5.11}$$

where $k + 1$ is the next higher quality supplied by the competitor. s is a market size parameter.

Given this set up, each firm is playing the following three-stage game. At stage one the firm decides on entry, at stage two it chooses quality and at stage three it competes *à la* Cournot in quantities or *à la* Bertrand on prices. The intensity of price competition at stage three crucially affects the quality choice at stage two. The intuition is that product differentiation is a device to reduce price competition. The more intense is price competition the more firms move away from each other in the quality space. Bonanno (1985) has shown for the two-firm case that with Cournot competition firms choose the same quality (minimum differentiation), whereas with Bertrand competition they differentiate the products. In the Bertrand case, price competition is such that profits would be zero if both firms were supplying the same quality. Both firms can increase profits by moving away from each other in the quality space. With Bertrand competition, the equilibrium number of firms depends on the income distribution: the broader the income distribution, the larger the number of firms that can coexist. For instance, it can be shown that if $2a < b < 4a$ only two firms enter. Assuming zero marginal costs, their profits are:

$$\Pi_2 = \frac{s(b - 2a)^2 (u_2 - u_1)}{9\, u_2} \tag{5.12}$$

$$\Pi_1 = \frac{s(b - a)^2 (u_2 - u_1)}{9\, u_1} \tag{5.13}$$

Notice also that $\Pi_2 > \Pi_1$ – i.e. the firm supplying a higher quality has higher profits. But – which is perhaps more important – market structure does not depend on market size. An increase in market size does not lead to additional entry, but to higher profits of incumbent firms.

5.3 Product differentiation strategies in mobile telecommunications

During the analogue technology period, the supply of mobile telecommunications services was mainly provided by monopolies. In the absence of competitive pressure, there were few incentives for product differentiation. However, during the digital phase and the opening up of the sector to competitive entry, the strategy space for product differentiation was expanded in several directions. Some of these are now discussed in some detail.

5.3.1 Coverage

The major attribute of a mobile telecommunications network is its *coverage*. Mobile telecommunications services are supplied using a grid of cells connected to each other. As new areas are covered, or as traffic increases within a given area, the only option left to an operator is to invest in additional cells, thus making the cellular technology subject to constant returns to scale.[6] As more spectrum is released and new licences attributed to entrants, it can be argued that the market will become more fragmented and prices need to be brought in line with costs. However, this conclusion should not be taken for granted since merger or exit of firms may occur for other reasons, as will be seen below. Calls can be made only if there is a cell covering the area the user is travelling through. Consumers, who are themselves mobile, evaluate coverage differently, according to the utility they derive from completed calls. However, even if users differ in the intensity of their preferences, given services provided by two networks of different size sold at the same price, it is likely that all such consumers would prefer to join the wider network. In other words, network coverage is a parameter for *vertical quality difference*. This is true so long as customers are mobile enough. On the other hand, if a customer is always located in a narrow area that is covered by all competing networks, she will not care about coverage but only about the price she has to pay. For such a customer, networks are providing homogeneous services.

If firms have ability to differentiate their product by coverage, this may have important implications for market structure. In particular, the quality feature of network coverage implies that 'natural' oligopolies could emerge in the industry. This is shown by Valletti (1999) in a duopoly model that delivers a series of results that apply directly to the mobile industry.[7] For instance, price competition can be relaxed by building networks of differing coverage. Differences in product characteristics are reflected in differences in prices and operators with national coverage will be relatively more expensive than operators with smaller coverage. The nature of price competition is also affected by a *minimum coverage requirement* set by the regulator. However, depending on the distribution of the WTP and

---

[6] There are two studies available on this topic, both for the USA. McKenzie and Small (1997) find some diseconomies of scale and constant returns to scale at best. On the other hand, Foreman and Beauvais (1999), employing a richer set of data (a panel of 100 different cellular market areas for three years, served by the firm GTE) find some mild scale economies. Notice that these are results on technology alone, since marketing and administration costs are not included.

[7] Network coverage is taken as an indicator in a vertical product differentiation model. One has also to take into account that network coverage not only affects the consumer's WTP for making calls, but also benefits in terms of being reachable for receiving calls over a wider area.

discount rates, maximum or minimum differentiation outcomes may occur. Valletti assumes price competition among firms, while in practice firms are constrained by capacity limitations. The scarcity of the spectrum of radio frequencies can be modelled using quantity rather than price competition, and it is well known that quality differentiation is milder the less tough price competition is. In practice, continuous improvements have permitted a more efficient use of the spectrum so that while a quantity (Cournot) model would be more appropriate for the past, in the future a price (Bertrand) model will be closer to reality. Nevertheless, the industry may remain concentrated because of its 'natural oligopoly' nature.

In practice, however it turned out that in Europe all firms tended to provide full nationwide coverage.[8] The scope for product differentiation in terms of coverage was thus limited. In other countries such as the USA, where there was no nationwide licence *per se*, all major firms tended to provide similar degrees of coverage. If coverage was not ensured by the firm's own network, it was provided through 'roaming' agreements at homogeneous prices, as will be seen later. Nevertheless coverage can be an important strategic variable during the initial roll out of networks. The crucial role played by coverage is confirmed, for instance, in the UK, as discussed in chapter 3. The two incumbent operators, BT Cellnet and Vodafone, reached full population coverage of both their digital and analogue networks very quickly, while cheaper PCN firms (Orange and One2One) at first followed a strategy of differentiation. In particular, One2One covered only 40 per cent of the population five years after having been awarded a licence. The PCN firms later had to revise their plans for the roll out of their networks: as firms succeeded in matching each other's coverage, resulting in a decrease of product differentiation, price competition intensified in the market. A similar pattern also emerged in the German market (Nattermann, 1999), where firms initially managed to relax price competition by product differentiation. It was more the reduction in the scope for differentiation, not the increased number of firms, that led to falling prices.

5.3.2 Pre-paid cards

During the analogue phase, mobile telecommunications customers typically had a subscription with a periodic (typically monthly) billing period. However, the switch to digital systems permitted greater flexibility in the administration of customer payments for services. The introduction of pre-paid subscription was a major product innovation in the mobile

---

[8] This typically entails population coverage of 95 per cent or more.

telecommunications sector. The main concept was that the subscriber paid in advance to receive a credit for a certain amount of traffic to be consumed within a certain time frame: there was no monthly fee to be paid. Moreover, upon expiry of the credit the card could still be used to receive incoming calls for a certain time period. Though there were no rental charge to be paid, pre-paid subscribers tended to have a cost per minute of calling time. However, one of the main attractive features for the user of the pre-paid card was the full control over cost.[9] For the card issuer, the advantage was the absence of credit risk because the telecommunications services had been paid for in advance. This permitted the attraction of a customer basis which would otherwise have been excluded because of poor creditworthiness.[10] This was particularly important in countries with less developed capital markets, where almost all transactions were on a cash basis.[11] The lessened credit risk, while ensuring service availability, also lowered the adoption barrier for user groups such as children.

An important pricing aspect of pre-paid cards is the length of validity of the card (and number) after purchase (or recharge). This is important as it sets a lower bound on the minimum amount a user has to spend within a certain time frame. Once that time frame has elapsed, any remaining credit is cancelled. For receiving incoming calls this time limit is sometimes slightly longer – i.e. it allows the user to receive calls, but not to make calls.[12] Several firms allow users to carry over unused credit to the next card. This eliminates the downside risk for the user in not using enough of the calling credit.

The first pre-paid cards were introduced in Germany and Switzerland in 1995. These cards were for one-time use only and could not be recharged. They were very expensive if compared with a traditional mobile subscription and were mainly designed to satisfy the needs of international business travellers rather than for developing the domestic market. The first

[9] This was not entirely true for RPP-based networks, such as those predominantly in the USA, where users also had to allocate some credit for incoming calls. This is also one of the reasons why such networks were late in introducing pre-paid cards. FCC (2003) reports that in the USA pre-paid subscriptions accounted for only 5 per cent of the total mobile telecommunications subscriber base, whereas in Europe typically more than two-thirds of subscribers were on a pre-paid basis.

[10] An investigation in Australia showed that 40 per cent of applicants for a digital mobile telecommunications subscription were refused because they could not meet the credit checks (OECD, 2000).

[11] In particular, when economic agents have little collateral that can be used to support applications for telecommunications services. In many developing countries the number of mobile subscribers in fact exceeds the number of fixed lines because of the possibility of pre-paid subscription schemes.

[12] This, of course, applies for countries with a CPP regime only. The firm has an interest in not cutting off a subscriber with an expired call credit as this subscriber may still be valuable in generating incoming traffic on which the firm receives interconnection payments.

pre-paid rechargeable cards were introduced by Telecom Portugal's mobile telecommunications subsidiary TMN in 1995 and by Telecom Italia Mobile (TIM) in 1996. In both cases, there was a rapid increase of subscribers for pre-paid cards: very soon more than half of the subscriber base was on pre-paid subscription. The introduction of a pre-paid card also led to a rapid increase in the overall penetration rate for mobile telecommunications. In the Nordic countries, which until the late 1990s were actually leading in terms of mobile telecommunications penetration rates, the introduction of pre-paid cards occurred relatively late. In Finland, the country with the highest penetration rate in the world, the first pre-paid cards were introduced only in mid-1998. A detailed investigation in Sweden has also shown that 15–20 per cent of pre-paid cards are not actually activated. Moreover, the credit on the cards is not always used to the full amount. In countries based on the RPP principle, such as USA and Canada, the pre-paid card is less attractive. Such cards were also introduced later in those countries and found less diffusion.

5.3.3 Customer lock-in

Mobile telecommunications service firms have different means to lock in a customer and increase her switching costs. At a general level, a typical lock-in is the choice of incompatible standards. The mobile firms could choose a mobile technology whose handsets were not compatible with the rivals' technology.[13] This was, however, either not possible because of mandatory standards or in fact rarely done by mobile firms because positive network effects would increase price competition. Firms have scope for locking in customers by contract. Mobile firms, to lower the entry cost to subscribers, often use handset subsidies, for instance, and this investment is recouped by requiring a minimum subscription period (typically one–two years). The recourse to minimum duration contracts has been made difficult by the emergence of pre-paid cards and other solutions had to be found to tie the customer – for example, by making the handset workable only with a given subscriber identification module (SIM), a chip card in the handset with the main customer data supplied by the mobile telecommunications service provider. Such a 'SIM lock' makes a handset usable only with a given SIM card. This is an enforcement mechanism in the context of subsidising the purchase of a handset with a pre-paid card. With a pre-paid card it would otherwise be difficult to pre-commit a subscriber to using a certain network only for a given period of time. Though the operators tried to justify the SIM lock as a safety device

---

[13] See chapter 3.

against loss and fraud, some regulatory authorities (e.g. Finland) see it as anti-competitive practice and therefore ban it. A ban on SIM locks thus deters handset subsidies with pre-paid cards combined with a minimum time of contract.

The telephone number constitutes another means for locking in customers. As long as each firm has its own dialling codes (usually a prefix of some numbers), changing the firm also requires a customer to change the telephone number. For some customers, such as lawyers or doctors, who hand their phone number out to a large potential caller base, the cost of changing the phone number is very costly. To prevent lock-in by the numbering system and to provide a check on market power of firms, many countries have introduced mobile number portability.[14] This means that in spite of swapping mobile telecommunications supplier, the customer can preserve the in number.

5.4 Empirical research on market behaviour

There is a relatively limited literature that empirically investigates the market conduct of firms in the mobile telecommunications sector. Most of the econometric research is based on the duopoly phase of the US market. For instance, Parker and Röller (1997) estimated the price–cost margins of the 305 non-overlapping duopoly franchises that made up the US market during the early analogue phase. The econometric model is based on the assumption of a homogeneous good supplied by two firms. Firms maximise the profit function with respect to quantities. q. $p(q)$ is the market price and c is the marginal cost of supplying the service. The first-order condition for firm i in market v at time t is:

$$\theta \frac{dp_{tv}}{dq_{itv}} q_{itv} + p_{tv} = c_{itv} \tag{5.14}$$

The parameter θ indicates the degree of collusion. If $\theta = 0$ then price is equal to marginal cost and there is perfect competition. $\theta = 1$ would indicate Cournot competition. Any $\theta > 1$ is interpreted as evidence for collusion. Parker and Röller consider the conduct parameter as a function of a set of market characteristic variables, among which are

---

[14] For instance, the EU's 1998 regulatory framework did not provide for mobile number portability (it provided it only for fixed line telecommunications). The 2002 regulatory framework instead provided for mobile number portability. The USA introduced mobile number portability in 2003. In this case, implementation is complicated as fixed and mobile networks have the same numbering base. Number portability thus needs to be across both fixed and mobile telecommunications networks.

cross-ownership and multimarket contact. Prices are found to be significantly above marginal costs and firms appear to charge higher prices than Cournot competition would warrant. The highest mark-ups are not found in the mobile subsidiaries of the incumbent RBOCs, but rather in independent firms. In markets where the independents face each other exclusively, there is actually evidence for cartel behaviour. Cross-ownership and multimarket contact are important factors in explaining non-competitive prices. Busse (2000) refined the research of the scope for multi-market contact to increase prices. Firms show a tendency to charge the same prices when colluding, and these prices are 7–10 per cent higher than they would have been in absence of collusion. It is thus suggested that multimarket contact facilitates collusion not only by enhancing the ability to punish, but by increasing firms' scope for price signalling and coordination.

Other findings from empirical studies are that regulation leads to higher prices. This is consistent with results obtained by Hausman (1997) and Shew (1994). Regulation is suspected to be a device that facilitates collusion, the regulator could have acted as a cartel board which made firms' strategies, an essential role for detecting deviation from agreed pricing behaviour and for enforcing cartel agreements. Duso (2000) has investigated these aspects further, by considering pricing behaviour and regulatory decisions as occurring simultaneously. This study benefits from the fact that regulatory regimes vary across franchises, with some franchises not being regulated at all and so this approach helps to overcome the problem of selectivity bias. Neglecting this problem leads to the typical result that prices in non-regulated markets are generally lower than in regulated ones, but the impact of regulation is not observed to be significant.[15] On the other hand, it can be observed that in non-regulated markets prices fall once they become regulated. By making the regulatory choice endogenous, Duso is able to show that firms can prevent certain markets from becoming regulated, in particular those where regulation has had the strongest impact on prices. Miravete and Röller (2003) have estimated the effect of two-part pricing schemes in the early analogue phase of the US mobile telecommunications industry. Although the US mobile telecommunications franchises were duopolies, the firm associated with the wireline licence typically entered first and tended also to be slightly more efficient and thus had a higher mark-up. The authors demonstrate that the ability to establish two-part tariff schemes increased welfare.

Nattermann (1999), by using a similar model, estimated the mark-ups in the German mobile telecommunications industry over the period 1986–97.

---

[15] This was, for instance, the finding of Keta Ruiz (1995).

This study found evidence for cooperative pricing with average mark-ups of 37 per cent. The study also covers effects from non-price competition and provides evidence that product differentiation between firms (mainly in terms of coverage) has decreased over time, coinciding with decreasing price–cost margins. Lower prices attracted new customer segments with higher demand elasticity, which further increased price competition. The initially high margins in the German market are attributed to restricted entry; the entry of the third firm, in particular, has led to a strong decline in margins.

5.5 Theoretical foundations for pricing in mobile telecommunications

Pricing in the telecommunications sector used to be regulated. With the liberalisation of the telecommunications sector price regulation was rolled back and increasingly left to market forces, especially in market segments where competition appears workable. There is the presumption that competitive forces will lead to prices reflecting the cost of providing services and inducing enhanced efficiencies in firms. This may not be feasible in all market segments of the mobile telecommunications industry, however network effects may prevent competitive forces from working in the traditional way in spite of open access to infrastructure. For instance, with fixed line telecommunications, competitive entry occurs mainly in upstream segments of the market, such as long-distance services, with the monopolist maintaining its market power on access to the customer (the 'last mile'). New entrants in the upstream segment therefore have to seek access to the downstream customers via the monopolist. The access price charged by the monopolist has a fundamental role in determining the final price charged to the customer by the new entrant. This setting has been referred to in the literature as the 'one-way access problem'.

Although the one-way access pricing problem remains an important benchmark for access pricing, this does not seem to reflect interactions in the mobile telecommunications industry. The appropriate framework is that of interconnection between networks in which each network has its own access to final customers. For instance, mobile firms have their own customers and a great many calls are between customers of fixed and mobile networks. Competing networks also comprising final customers thus changed the issue to a two-way access problem. This led to two sorts of pricing decisions for each firm: the pricing of on-net calls, where in principle all cost elements were under control of the firm, and the pricing of off-net calls where interconnections payments were due. Each case had different implications for price setting and regulatory requirements. As will be seen, competition among networks may not necessarily produce efficient

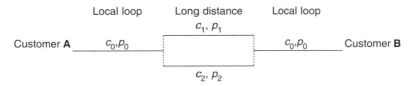

Figure 5.1 *Stylised representation of competition in long-distance telecommunications services*

outcomes. For on-net pricing it is assumed that competition among firms induces price competition, which can be mitigated in several ways by strategies of product differentiation or cooperative pricing, as seen above. For the analysis of off-pricing interconnection issues also have to be taken on board. Although in the mobile telecommunications industry the two-way access model is the more pertinent one, the one-way access model is important as a benchmark. Both will thus now be dealt with in some detail.

5.5.1 The one-way access problem

The standard access pricing problem concerns an upstream firm that controls a bottleneck facility and thus wants to set a price for the access to the facility to recover a monopoly profit. High access prices either prevent entry or maintain monopoly-level retail prices. To favour competitive entry, access price should be set low. But if the access price is too low, entry of inefficient firms may occur and incentives for ensuring long-term efficiency of the bottleneck facility may be insufficient. The theoretical literature on establishing the appropriate access price for this 'one-way access' problem has grown quite large. The general conclusion is that regulation is needed to avoid anti-competitive behaviour.[16] A simplified formal presentation of the issues is now made.[17]

Consider the market for telecommunications services between customers **A** and **B**, as indicated in figure 5.1, which represents a stylised network. Assume there are two firms and two market segments, one of which is competitive (long-distance telecommunications services) and one of which is a natural monopoly (fixed line local loop). One firm, the incumbent, is in both markets, whereas the other firm is in the competitive market only. Let c_1 be the marginal cost of production in the competitive segment of the market of the incumbent. c_2 is the marginal cost of the incumbent for the same service. c_0 is the incumbent's unit marginal cost in the bottleneck segment for originating or terminating a call.

[16] See Laffont and Tirole (2000) and Armstrong (2002) for surveys of these issues.
[17] This presentation relies on Laffont and Tirole (2000).

Let k_0 be the sunk cost for investments in the local loop infrastructure and a the access charge the entrant pays for accessing the subscriber in the local loop. q_0 is the total number of local calls, and q_1 and q_2 are the number of long-distance calls for the incumbent and the new entrant, respectively. The total number of calls is thus $Q = q_0 + q_1 + q_2$. The total cost function for local loop services is $C_0 = 2c_0Q + k_0$. The total cost for long-distance services for each firm is given by $C_i = c_iq_i$, for $i = 1, 2$. p_0 is the price of a local call and p_1 and p_2 are the prices for long-distance calls charged by the incumbent and the new entrant, respectively. The industry profit is:

$$\Pi(p_0, p_1, p_2) = (p_0 - 2c_0)q_0 + (p_1 - c_1 - 2c_0)q_1$$
$$+ (p_2 - c_2 - 2c_0)q_2 - k_0. \tag{5.15}$$

$S_0(p_0)$ is the consumer net surplus for local calls and $S(p_1, p_2)$ is the consumer net surplus for long-distance calls. With these assumptions, several pricing problems can be formulated, depending on the objective function to be maximised. The literature typically indicates three pricing rules: social welfare maximising prices, efficient component prices and cost-based prices. They are derived as follows.

Social welfare maximising prices (Ramsey prices)

The problem of socially optimum access pricing can be formulated as the prices that maximise the following social welfare function:

$$\max \{S_0(p_0) + S(p_1, p_2) + \Pi(p_0, p_1, p_2)\}$$

$$\text{subject to } \Pi(p_0, p_1, p_2) \geq 0 \tag{5.16}$$

With fixed costs equal to zero ($k_0 = 0$) price–cost margins are zero as well. With positive fixed costs ($k_0 > 0$) the price–cost margin is positive, and given by:

$$m_i = \frac{\lambda}{1 + \lambda} \frac{1}{\eta_i} \quad \text{for } i = 1, 2, 3 \tag{5.17}$$

The price–cost margin for service i is a function of the shadow cost of the budget constraint, as indicated by the Lagrange multiplier λ, and by an appropriately defined demand elasticity η_i that takes into account possible substitution and complementarities among goods. From this the optimal access price (Ramsey access charge) can be derived as:

$$a = 2c_0 + \frac{\lambda}{1 + \lambda} \frac{p_2}{\eta_i} \tag{5.18}$$

The socially optimum access price is higher than marginal access cost and requires long-distance services to contribute to fixed costs in the local loop.

This price is also cost and usage-based and decreases with demand elasticity. Given the large information requirement, Ramsey prices are little used in practice, as it would be difficult to calculate them. One response to this could be that pricing decisions should be delegated to those that have the necessary information (i.e. the firm itself), using a *global price cap* on the entire range of services supplied by the firm.

Efficient component pricing rule

There are various definitions of the efficient component pricing rule (ECPR) found in the literature and they have generated substantial controversies (see Armstrong, 2002). One formulation claims that the incumbent should not set access prices above its opportunity cost on the competitive market segment (i.e. $a \leq p_1 - c_1$). The ECPR satisfies the socially optimum Ramsey pricing rule only under very specific circumstances. However, proponents of the ECPR as a regulatory rule have emphasised the relatively limited information requirement which makes the principle useful for practical purposes. The ECPR implies that only entrants with a genuine cost advantage over the incumbent can enter and entry is in principle neutral to the profits of the incumbent.

Cost-based prices

Cost-based prices can be defined in two ways: backward looking prices are based on historical costs for the sunk cost element; forward-looking prices are based on current estimates of the sunk costs. In a context of rapid technical change, these two cost definitions may differ substantially.

Backward looking prices There are two forms of backward looking prices, depending on whether mark-ups are additive or proportional. With additive mark-ups, a mark-up that is usage-proportional equal to k_0/Q is added to the price of each service. The access price becomes:

$$a = 2c_0 + \frac{k_0}{Q} \tag{5.19}$$

The access price thus determined satisfies also the ECPR since $a = p_1 - c_1$.

With proportional mark-ups the marginal cost of each service is multiplied by a factor, z. The access price is thus $a = z2c_0$. $z > 1$ is chosen such that the sum of the mark-ups becomes the fixed cost k_0, or:

$$(z - 1)[(2c_0)q_0 + (2c_0 + p_1)q_1 + (2c_0 + p_2)q_2] = k_0 \tag{5.20}$$

The access price becomes $a = p_1 - c_1 - (z - 1)c_1 < p_1 - c_1$, which means that it is below the ECPR. In other words, cost recovery falls more than proportionally on the competitive segment. This could lead to inefficient entry or inefficient bypass.

The benefit of fully distributed prices, an indicated, is that it does not lead to 'regulatory takings'[18] – i.e. the incumbent is always allowed to recoup its past investment. This may not be so with forward looking cost pricing.

Forward looking cost-based prices

With forward looking cost-based prices, the access price is based on the lowest cost of the asset at a given time. In this sense, the long-run incremental cost (LRIC) is a measure of the economic cost of an asset. It is based on the assets being valued and depreciated on a current cost accounting (CCA) basis, giving the current replacement cost of a modern efficient asset. The determination of the access price is thus *time sensitive*, particularly in the telecommunications industry where rapid technological change relentlessly reduces the price of equipment and increases its performances. The access price at time t can be expressed as a function of the interest, r, the rate of technological progress, x, and the rate of physical depreciation, δ, yielding an access price a at time t in the following way:

$$a_t = (r + x + \delta)c_t \qquad (5.21)$$

Regulatory takings occur when the incumbent is forced to lower its access price to the LRIC of the new and more efficient technology, and thus to lower the access price below its own LRIC. To break even, the incumbent has to add a mark-up on its current LRIC equivalent to the rate of the technological progress, x.

5.5.2 The two-way access problem

Whereas with the one-way access problem only the owner of the bottleneck facility (for instance, the local loop) has direct access to customers, with the two-way access case the firm seeking access also has direct access to its own subscribers. In other words, in the 'two-way access' case the two networks have to be interconnected in order to give access to each other's subscribers. Two distinct two-way access situations can emerge, one without and one with competition for customers. The first situation is familiar from international telecommunications where networks are located in different countries and there is no competition for the subscriber base. The interconnection arrangements for international calls are regulated by the international accounting rate regime.[19] The second situation is typical of

---

[18] This term was coined by Sidak and Spulber (1996).
[19] See Wright (1999) on the principles of this system and the problems of relating this to cost-based prices.

competitive domestic fixed line telecommunications (CLEC) or mobile telecommunications where firms fight for customers.

Within the context of telecommunications networks, two parts of the service can be distinguished: originating the call and forwarding it to the point of interconnection and the termination service, which involves picking-up of the call from the interconnection point and transferring it to the final destination. With telecommunications, the entire call is in most instances paid for by the originating party of the call (i.e. the CPP principle). There may also be cases where the receiving party pays for the termination services (i.e. the RPP principle), as will be seen shortly.[20]

Calling party pays (CPP)

Assume two mobile telecommunications firms, A and B, and one fixed line firm, C. For reasons of simplicity assume that all mobile customers call fixed line subscribers. The demand for calls from fixed line subscribers to any of the two mobile networks $i (i = A, B)$ is $q_{ci} = q_{ci}(p_{ci}, q_i)$, where p_{ci} is the price charged by network C for calls to network i. A particular assumption is made concerning the effect of positive network externalities – i.e. the demand for calls from network C to network i increase with the demand for network i services $(dq_{ci}/dq_i > 0)$. Assume linear costs, with c_i the access cost of network i and a_i the access charge that network i charges (for both call termination and call origination). It is assumed that mobile networks have higher access costs – i.e. $c_i > c_c$ for $i = A, B$.

For each mobile network $i = A, B$ the optimisation problem consists in maximising the following profit function with respect to service and access prices:

$$\max_{p_i, \, a_i} \Pi_i = (p_i - a_c - c_i) \, q_i \, (p_A, p_B)$$
$$+ (a_i - c_i) \, q_{ci}(p_{ci}, q_i(p_A, p_B)) \tag{5.22}$$

The first part of the sum concerns the profit from off-net calls (i.e. from calls to network C) and the second part indicates the profits from incoming calls. From the latter, it is clear that each mobile network will set access charges above access costs (i.e. $a_i > c_i$). Notice also that the choice of setting the retail price, p_i, also affects the incoming traffic volume, q_i.

For the fixed network C the optimisation problem consists in maximising the following profit function with respect to service and access prices:

$$\max_{p_{Ci}, \, a_C} \Pi_C = (a_C - c_C)(q_A + q_B)$$
$$+ \sum_i (p_{Ci} - a_i - c_C) q_{Ci}(p_{Ci}, q_i) \tag{5.23}$$

[20] The following analytical presentation follows Doyle and Smith (1998).

The following observations can be made about the first-order conditions for profit maximisation of the above two equations:

1. Each mobile firm as well as the fixed line firms sets $a_i > c_i$ for $i = A, B, C$ (i.e. the access price is above access cost).
2. $p_{Ci} > a_i + c_C$ (i.e. the fixed network sets the price for calls above the costs associated with such calls). Thus the higher the access price, a_i, the higher is the price for the call (double marginalisation).
3. The retail price, p_i, is a function of the rival's price and the price set by the fixed network.
4. It is possible that the mobile firm may set its price below cost (i.e. $p_i - a_C - c_i < 0$ for $i = A, B$) as long as this attracts additional customers. Losses are recouped with high enough access margins ($a_i - c_i$) (double marginalisation).
5. The price for fixed to mobile services is in a positive relationship with the price set for the mobile to fixed market.

Receiving party pays (RPP)

Assume that the fixed line firms set the price for fixed to mobile services at the same level as fixed to fixed lines services (i.e. at c_c) and charges the same for access to its network. The 'posted' price for fixed to mobile calls is $p_{Ci} = a_i + c_C$. The customers of the mobile network are assumed to pay $p_{Ci} - c_C$ for receiving a call. The demand function for each mobile network firm has changed, and each mobile firm maximises the following:

$$\max_{p_i, \, a_i, \, p_{Ci}} \Pi = (p_i - c_c - c_i) \, q_i \, (p_A, p_B, p_C)$$
$$+ (a_i - c_i) \, q_{ci}(p_{Ci}, q_i(p_A, p_B, p_C)) \tag{5.24}$$

Unlike the CPP case, the demand for mobile to fixed services now depends on the posted price, p_{Ci}. The problem can be simplified as being the choice of a_i that determines also the posted price, p_{Ci}. From the first-order conditions it emerges that the expressions ($p_i - c_c - c_i$) and ($a_i - c_i$) are positive. The higher the elasticity of calls from fixed to mobile customers as a response to increased demand for mobile services, the closer access charges are to marginal cost. Imperfect competition, however, leads to prices above access cost. Access charges are more above incremental cost the higher is the responsiveness of demand for fixed to mobile calls to the demand for mobile services.

Under RPP and with prices for fixed network services equal to cost (by means of regulation or competition), the elimination of double marginalisation lowers prices. Access charges set by the mobile networks could be lower than in the CPP case if the demand for the net mobile network's services is more elastic with respect to access charges. RPP has also the advantage of avoiding the need for regulatory intervention imposing uniform fixed to mobile charges in order to avoid discrimination.

5.6 Welfare analysis of charging regimes

Generally both parties (the calling party and the receiving party) derive some *utility* from the call, though these utilities may differ. For fixed telecommunications networks in all countries and for mobile telecommunications in most countries there is the principle that the calling party pays for the call (CPP). This set up creates a *call externality*, as the receiver derives a utility that is entirely paid by the calling party. Inefficiency derives from the fact that the calling party has also to pay for the cost of terminating the call. This induces too low a number of originating calls because, at equilibrium, the marginal utility of the calling party is higher than the cost of setting up the call and the marginal utility of the receiving party is lower than the cost of receiving the call. Apart from inefficient traffic patterns, additional distributional issues arise when different networks are involved.

With CPP, revenues are levied in the originating network and the terminating network has to be compensated for terminating the call. This requires the establishment of interconnection regimes where terminating networks receive a fee from the originating network. Because the calling party pays the entire cost of a call, the cost charged to the calling party also includes an allocation of common costs from the receiving party network in the form of termination fees. Cost-based fees, such as the LRIC, are second-best benchmarks for termination fees and do not provide incentives for cost efficiency in the receiving party network. Indeed, the termination fee of the receiving party network usually does not matter in the decision of the call originator becoming a customer of a given network.

There is an open theoretical debate on how the cost of the call should be distributed between the two parties.[21] For instance, DeGraba (2002) compares the telephone call to the joint consumption of a public good by the two parties. In that case, the Lindahl equilibrium – i.e. the optimum quantity of call minutes – is achieved when the sum of the marginal benefit of each customer of a public good equals the marginal cost of producing that unit. The implication would be zero interconnection charges and each network would adopt a bill-and-keep practice (DeGraba, 2000). As a consequence, the receiving party would be charged for the call as well.

---

[21] For a survey of the different approaches, see Taylor (1994).

Atkinson and Barnekov (2000) argues that bill-and-keep regimes are competitively neutral (i.e. they do not favour a particular technology or firm), economically efficient and require minimum regulatory intervention. However, such simple principles of economic efficiency are sometimes in conflict with practical implementation, as there are often legacy interconnection systems in place. Changes in these have strong distributional implications: a particular interconnection policy not only distributes the costs of calls among users, but also pursues other goals such as promoting competition and correcting network externalities. The resilience of the CPP principle has been imputed to two facts: first, traffic patterns are generally balanced (i.e. a user makes on average as many calls as she receives) and thus calling externalities cancel out; second, in the early days of telecommunications, technical and administrative costs for also collecting receiving party fees would have been too high to justify the system. Both premises, however, are nowadays no longer more applicable as digital technology has led to low costs in billing and the proliferation of new networks has led in many instances to imbalanced traffic.

In spite of all the advantages in terms of inducing economic efficiency, RPP is rarely used. Where it is, the main reasons are other than those of economic efficiency. For instance, RPP use in the US mobile telecommunications market is due to the fact mobile service subscribers do not have different codes from fixed line subscribers and thus the caller cannot see from the number whether she is dialling into a fixed or mobile system. As call charges to mobile systems are typically much higher than to fixed systems, it was considered as unfair to leave the caller uncertain whether she is calling a fixed or a mobile telephone number.

Whereas it is uncontroversial that the calling party has to contribute to the cost, as it is the party most likely to derive utility from the call, it is much more debatable to what extent the receiving party should contribute. A significant problem in introducing RPP is the possibility of nuisance calls that give negative utility to the receivers. This could be mitigated, for instance, by allowing a few free seconds for the receiving party to decide whether to continue with the call or not. Another problem is the reluctance of users to receive calls if charges are high. This is one of the reasons why US mobile users tend to be reluctant to advertise their mobile phone number, or switch the handset off when not phoning.

The fact that there is generally some form of competition among firms in the mobile telecommunications market has induced a less severe regulatory regime than for fixed line telecommunications. The justification is that competition of customers is sufficient to constrain retail prices. Nevertheless, there are some segments of the market where firms have market power and this is translated in high prices – in particular, high termination charges for mobile firms. There is now an extensive theoretical literature[22] that has analysed the welfare implications of high termination rates. For instance, Wright (2000, 2002) shows in a vertical product differentiation model that firms set termination charges above cost, provided market penetration has not reached 100 per cent. Termination revenue becomes a substitute for the retail revenue that is directly paid by customers. This gives scope for lowering retail prices and hence for increasing the number of mobile customers. Only when 100 per cent penetration has been achieved in the mobile telecommunications market do the incentives for mobile firms to set termination charges above cost cease. With this framework, it can also be shown that this transfer to the mobile sector through high termination charges can be efficiency enhancing. This can happen because of the following two externalities. First, the penetration of the mobile market is held back because of imperfect competition. The subsidy from the fixed network allows mobile firms to lower prices and hence drive penetration further. Secondly, there is a positive network externality through the increased number of mobile users, and this is not priced in by mobile firms.

In theory, competition among networks may not be needed to avoid anti-competitive behaviour (foreclosure), but access pricing may nevertheless be used as a device for collusive behaviour. As shown by Armstrong (1998, 2002), high access charges increase the cost of reducing retail prices unilaterally because by lowering the price the network causes an increase in off-net calls which incur high access charges. Seen another way, the networks benefit from setting the access charge above the marginal cost of access because this mutually raises the rivals' cost and generates higher profits without collusion in the retail market. Nevertheless, as Laffont and Tirole (2000) show, there are reasons why high access charges may not facilitate collusion – high substitutability of services or termination-based price discrimination (i.e. different prices for off-net and on-net calls), for example.

There are thus no clear-cut theoretical predictions on the degree of competition between networks and cost-based pricing. Everything depends on the parameters of the systems. The relationship between social welfare maximising access price and the marginal cost of access depends on the existence of joint and common costs. In the absence of these costs, the socially optimal termination charge lies below the marginal cost of access. The reason is that the market power of firms enables them to set retail prices above marginal cost and these profits are set off by negative access margins. In the presence of joint costs, the socially optimal termination

---

[22] See Armstrong (1998, 2002) and Wright (2002).

charge may lie below or above the marginal cost of access, depending on the size of these joint costs.[23] One may thus conclude that competition among networks is neither a necessary nor a sufficient condition to avoid anti-competitive behaviour.

Neither is the switch from CPP to RPP necessarily welfare improving. Kim and Lim (2001) show that calling prices with RPP are lower than with CPP, profits are higher with RPP than with CPP but the outcomes for social welfare are ambiguous. Jeon, Laffont and Tirole (2001) provide a justification of RPP based on the fact that receivers derive a utility from receiving calls (call externality) and that they can affect traffic volume by hanging up (receiver sovereignty). They find that with regulation of both reception and termination charges an efficient equilibrium can be attained, with termination charges below termination costs. With unregulated reception charges, it is optimal for each network to set prices for emission and reception with their off-net costs, and for appropriately chosen termination charges the equilibrium is efficient. However, network-based price discrimination creates strong incentives for connectivity breakdowns. With the CPP principle unregulated mobile termination charges tend to be at too high a level, whereas with RPP it is questionable whether an unregulated equilibrium exists at all (Armstrong, 2002). The general conclusion of the literature is that there may be little scope for relaxing sector specific regulations.

5.7 Mobile telecommunications pricing, by type of service

This section divides the pricing analysis into two large segments: wholesale services and retail services.[24] The *retail mobile market* refers to goods and services sold directly to end users – i.e. the provision of access and national and international mobile telephony and 'roaming' services to end users. *Wholesale services* describe the provision of any kind of service by a mobile network firm to another party – e.g. a service provider – who incorporates this service into its own retail services supplied to end users. The wholesale mobile sector covers those mobile services used as inputs to products or services sold in retail markets. The termination of incoming calls to mobile networks is thus a wholesale service sold to the originating operator that sells the incoming call to its fixed or mobile retail customer. Wholesale services include the provision of wholesale access and capacity/airtime to national and international 'roaming' partners.

---

[23] See Laffont and Tirole (2000) for an analytical derivation of these propositions.
[24] For a detailed description of market definitions, see Oftel (2001a).

5.7.1 Retail prices

There are two segments in the retail service market: *access* and *airtime or usage*. From the consumer's point of view, access can be thought of as the ability to make and receive calls. From the perspective of the network firm, this involves registration and recognition by the network as well as the supply of a SIM card. As with fixed telecommunications services, three basic retail services can be identified for which each customer may be charged separately – connection, subscription (rental) and call charges (usage). A variety of non-linear pricing schemes can be designed. Of course, there can also be additional services, or one or more of these charges can be waived, or several services may be offered as a bundle. For instance, with pre-paid cards there is no monthly subscription, but there are minimum usage requirements. We now briefly describe some of these services.

Connection

An up-front fee is usually charged for connecting new subscribers to the network. The reason for this is to compensate for the costs the operator incurs to register a new subscriber and for the SIM card. In the early days of the industry, these charges used to be very high, but with increasing competition for new subscribers fees have declined significantly and are nowadays waived in several cases, or even negative (e.g. through promotional activities such as handset subsidies).

Subscription

With post-paid subscription, a fee is typically charged on a monthly basis for access to the moblie network. The charge may in some cases include a certain number of free minutes of airtime. Subscription charges are typically part of a two-part tariff scheme together with calling charges, whereby a high (low) subscription charge is coupled with a low (high) calling charge.

Airtime

Air charges are usually based on a per-minute charge. With CPP, airtime is normally charged only for outgoing calls, unless the receiving party is 'roaming'. With RPP, airtime is also charged for incoming calls. Prices are steadily declining; in addition, price competition in many countries is reducing the metering unit from minutes to seconds. This allows users to be charged more precisely according to their effective calling time, and acts as a further incentive for lower prices. Calling charges are differentiated by time of the day (peak vs. off-peak) and by destination (on-net vs. off-net).

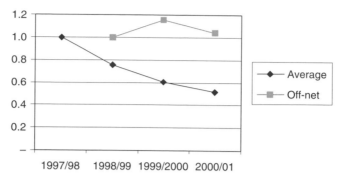

Figure 5.2 *Evolution of mobile pricing, UK, 1997–2001*
Note: The figure indicates the evolution of pricing for off-net mobile calls compared with the average mobile airtime price in the UK, running from April to March of the following year (1997/98 = 1).
Source: Oftel.

In Europe, where mobile licences are generally granted at a nationwide level, domestic calls are typically not charged by distance. Nevertheless, some firms have tried to introduce so-called 'home-zone' tariffs where the user is charged less for calls placed within a certain area. This tariff scheme was adopted to encourage the possible substitution of fixed for mobile subscription; indeed, the difference between fixed and mobile tariffs is much less for local calls than for long-distance calls. In the USA, an opposite trend can be observed: initially, airtime charging was not made at uniform national rates because firms had regional licences. However, with the emergence of firms with nationwide coverage uniform nationwide tariffs have begun to emerge.

Although with CPP national airtime pricing may not be distance-sensitive, it certainly is sensitive to the *destination network*. In general, one has to distinguish calls made within the same mobile network (on-net calls) and calls to other networks (off-net calls), which can be on either mobile or fixed networks. The tariffs applied can also be different according to the network, because the cost bases for the tariff calculations may be different. On-net call costs are entirely internalised by the firm, whereas for off-net calls the firm has to pay access charges which are typically above cost. This is the reason why prices for off-net calls have fallen much less than prices for on-net calls. Figure 5.2 shows the evolution of pricing for off-net mobile calls compared with the average mobile airtime price in the UK. This reflects a more generally observed trend that, until the emergence of more than two firms in the industry, firms used to typically price all out-going mobile voice calls (that is, on-net mobile to mobile, off-net mobile to mobile and calls to fixed telephones) identically. With further entry, mobile operators halted reductions in the prices of off-net mobile calls. As a result, off-net mobile call prices turned out to be relatively high compared with other domestic voice calls. This may be explained by the structure of the price of an off-net mobile call. The price paid by a subscriber is the value of the wholesale termination rate levied by the network being called, plus the value of the retention levied by the subscriber's own network. Mobile termination rates are substantially higher than the termination rates set by fixed networks. It is likely that this difference in the termination rate has helped to prevent the price of off-net calls from falling to the level of other mobile outgoing calls. Oftel (2001a) estimated that an originating mobile firm retained more for a call to another mobile network than for one to a fixed network. At the same time, there is no evidence that competition is acting to reduce this retention.

Strategic considerations on network externalities have to be taken into account in the pricing of domestic mobile to mobile (MTM) calls. It may be rational for the network with the highest market share of subscribers to maintain high off-net prices and low on-net prices because the customers of large networks are more likely (relative to the customers of a small network) to make more on-net calls. It seems reasonable that a consumer is more likely to be attracted by low on-net call prices offered by a large network than by a small one. On the other hand, because of the limited ability of a smaller network to reduce its off-net call prices because of the relatively high termination charge for off-net calls, a rational response of a small network to a large network's strategy of setting relatively high off-net calls consists in simply pricing off-net calls at a similar level to the larger network. However, because there is little evidence for differences in demand elasticity, having significantly different on-net and off-net prices is very inefficient from a social welfare point of view.

Concerning calls from mobile to fixed (MTF) networks, mobile firms typically set uniform prices for national calls, irrespective of whether the call is local or long-distance. Prices for MTF calls have been declining, but less than prices for long-distance calls within the fixed network. Mobile call tariffs can sometimes be lower than national long-distance calls. For local call prices, a different trend applies. Because in most countries local calls used to carry a low price (typically local calls were cross-subsidised by long-distance calls), there was little leeway for reducing local call prices further. In the RPP context, many of these problems do not arise as the mobile customer pays for the origination of the call but not for the termination. As the FCC (2003) reports, in the USA the average price per minute was $0.12 compared to prices typically in excess of $0.20 for European countries which have CPP.

5.7.2 Wholesale prices

'Roaming'

When a cellular subscriber uses the phone outside an area covered by the network of its mobile firm she is said to be 'roaming'. 'Roaming' services can therefore be thought of as providing temporary access to customers from other networks. There are two different types of 'roaming', depending on the origin of customers: national and international. *National* 'roaming' occurs when a subscriber uses her phone within her own country, but at a location which is not covered by the network of her mobile firm. This subscriber therefore uses the network of another, in principle competing, firm. National 'roaming' among holders of a mobile telecommunications infrastructure licence is normally restricted because of concerns about anti-competitive behaviour. It is, for instance, temporarily allowed for new entrants while they are setting up their own infrastructure. The availability of national 'roaming' reduces the entry cost of new firms, as they are able to provide full national coverage without having a national network. Incumbent firms may refuse to grant national 'roaming' or charge high prices to make it unprofitable. In this case, the regulator may be invoked or inefficient traffic patterns may be induced.[25]

As mentioned, national roaming in most countries is just a temporary device for easing the entry of new operators. There are only a limited number of countries where national 'roaming' is a permanent arrangement, countries where operators do not have a nationwide licence. However, in those cases there is a trend toward uniform 'roaming' charges.[26] National 'roaming' is important for mobile telecommunications firms without an own network, the so-called mobile virtual network operators (MVNOs). MVNOs buy bulk airtime from facility-based firms and resell it to retail customers, possibly bundling it with other service elements.

International 'roaming' occurs when a cellular subscriber is abroad and uses the network of foreign firms. There is an additional feature for international 'roaming' for subscribers in a home network practising CPP. For all inbound calls, the mobile subscriber 'roaming' abroad has to pay a fee, which means that the subscriber is therefore subject to the RPP principle for international 'roaming'.

---

[25] This happened in Finland where the newly entering firm Telia asked for national 'roaming' to existing operators. As this was not granted, Telia accessed the incumbent's mobile network using Swisscom's 'roaming' agreement. These calls were thus filed as international calls, though they were in fact mostly national. A similar agreement was made with the fourth German mobile firm Viag Interkom (OECD, 2000).

[26] See OECD (2000).

The pricing of 'roaming' services is undertaken in a two-step procedure. The first is *wholesale pricing*, where the visited network firm charges the user's home network firm for the calls, usually by adding a mark-up of 15 per cent on the normal network tariff. The second stage is *retail pricing*, where the home network firm sets the price for the subscriber, normally by adding a 10–35 per cent margin on top of the wholesale price set in the first stage. The 'roaming' price to the final customer thus obtained is therefore the result of double marginalisation. Technical constraints make it difficult to escape this mechanism: visited mobile firms have no incentive to lower the wholesale prices because their counterparts cannot guarantee traffic redirection to their network. This inability to direct traffic to the cheapest parallel network in a visited country makes it difficult to apply competitive pressure. Moreover, mobile users typically show ignorance about the 'roaming' costs charged by visited networks[27] and this induces an absence of price competition at the retail level.

It is therefore not surprising that the institutional setting for fixing prices for international 'roaming' services have attracted attention of regulators and competition authorities, especially in Europe.[28] During the early days of GSM, which introduced international 'roaming' on a large scale, the wholesale 'roaming' charges were based on the 'normal network tariff' (NNT). The NNT was defined as the basic standard user tariff applied to the majority of users. Over time an increasing mark-up was placed on the NNT, reaching the maximum permitted by regulators of 15 per cent in 1995. Moreover, NNTs were fixed over longer time frameworks and therefore did not respond quickly to reductions in domestic retail tariffs. The NNT was later abandoned for the inter operator tariff (IOT), by which mobile operators apply a negotiated wholesale tariff to their 'roaming' partners. Although in theory this should have increased the freedom of mobile network operators to compete directly on wholesale 'roaming' rates, there is no evidence that such competition has emerged. On the contrary, a comparison between NNT-based prices and IOT prices at the end of 2000 showed price premia in excess of 200 per cent for certain services.[29] One of the reasons is the fact that the IOT regime has cut the link with the price of non-'roamed' retail calls. This has made wholesale 'roaming' prices resistant against the competitive pressure that could have come from the reduction of prices in non-'roamed' calls Europe-wide. Most operators charged calls at a higher IOT for European 'roamed' than for domestic 'roamed' calls. There is also high price variation across countries.

[27] See Oftel (2001b).   [28] See European Commission (2001).
[29] See European Commission (2000).

Interconnection

Interconnection pricing concerns the price of services of a given network to other networks for access to the network's customer. In the early days of the industry, mobile telecommunications firms in most countries had a special regime for determining their access prices. In principle they were not bound to have cost-based prices and large asymmetries in interconnection charges could occur between fixed and mobile networks. A mobile network is in a monopoly position for providing access to any of is customers. As seen already, high interconnection charges are a device for collusion. Only with the increase of mobile networks did the regulator start to insist on cost-based prices for interconnection, though strong asymmetries still persisted and were justified by the higher operating costs that new mobile networks had compared to older fixed line networks. According to the EU regulations, only mobile firms with the status of 'significant market power' as defined by the national regulator had to publish a non-discriminatory cost-based interconnection offering.

Fixed to mobile (FTM) termination

The access charges in the case of FTM calls is a special case. Armstrong (2002) refers to this as the 'competitive bottleneck'. There are several networks competing vigorously for the same pool of subscribers, but these networks have a monopoly position in providing services to their subscribers. The termination charge set by a mobile firm does not directly affect the customers of its own network, but does affect the price set for the fixed line subscriber. As seen already, the market power of the mobile firm provides the incentive for setting termination charges above termination costs. This extra profit can then be used for subsidising the acquisition of new subscribers and locking them in through devices such as handset subsidies. Mobile firms offer new subscribers handsets below cost in exchange for a minimum subscription period; however, in this case the cost of the handset subsidy is not covered by the airtime payments of the mobile subscriber, but by the fixed line subscribers calling the mobile network. Mobile subscribers benefit from this pricing policy, but their benefits are more than outweighed by the cost it imposes on fixed line subscribers. The rapid expansion of the mobile subscriber base is therefore being subsidised by the fixed line network (Wright, 2002). High FTM prices induce an inefficiently low traffic from FTM networks. To redress this situation, regulatory action is required to lower mobile termination charges to marginal cost (Gans and King, 2000; Armstrong, 2002) and possibly to increase mobile retail charges to cover for common and fixed costs (Hausman, 2000). However, there is still an ongoing debate in the literature as to whether FTM calls should be regulated or whether there are countervailing mechanisms that may relax the requirement for regulation. Such mechanisms might be information about and internalisation of the MTF prices in the decision to join a mobile network.

Part of the market power of the mobile operators is caused by consumer ignorance regarding internetwork pricing. Empirical evidence[30] shows that a fixed line subscriber, when placing a call to a mobile subscriber, has little knowledge about which network will terminate the call and what will be the price. This could be deduced *a priori* from the telephone number, but this possibility disappears with number portability where the network is no longer unambiguously determined by the telephone number. Measures that increase information to fixed line subscribers should thus have the effect of reducing termination charges (Gans and King, 2000).

The internalisation of MTF prices in the decision to join a mobile network could essentially work in two ways. The first could be by allowing subscribers to care about their callers, which could be the case where regular callers were family members or close friends. This would provide incentives for the mobile user to subscribe to networks with lower termination costs. The second could be by charging mobile users for incoming calls (i.e. switching to a RPP regime). In that case, calls from the fixed to the mobile network would be charged at regular fixed call rates to the fixed line caller and the mobile recipient of the call would pay for the termination on the mobile network.[31]

The key regulatory question is who fixes the FTM price. In practice, the price is generally determined by the power of the negotiating parties involved. In some countries with CPP (e.g. France and Portugal), it is the mobile firm that sets the FTM charge, which means that it sets the prices to be charged to the fixed line users. The advantage of this system is that in presence of more than one operator an element of competition in the setting of FTM prices is created. In other countries, it is the fixed line firm that sets the FTM price and the payment to the mobile firm is settled through an interconnection agreement. There is also the possibility that fixed and mobile firms can agree jointly on FTM call prices and thus different FTM call prices may exist depending on the mobile network to which the call is directed. In practice, mobile firms typically retain a far higher share of the retail price of a FTM call than fixed firms.

---

[30] MMC (1998)
[31] See Doyle and Smith (1998) and DeGraba (2000) for more details.

5.7.3 Mobile virtual network operators

Until now, the market structure has been analysed with reference to firms that have a spectrum licence. However, there are also other firms that can supply mobile telecommunications services. Mobile virtual network operators (MVNOs) are suppliers of mobile telecommunications services to retail customers; there is no precise definition of an MVNO, but the most important characteristic is not having direct access to radio spectrum and therefore requiring a contractual relationship with one or more firms with a spectrum licence. A MVNO typically supplies the customer with a SIM card and has full control over the contractual relationship with the retail customer. Although MVNOs may in some certain aspects resemble the service providers known from the early days of the mobile telecommunications market,[32] there are nevertheless important differences.[33] MVNOs buy very large amounts of traffic minutes from firms with a spectrum licence at cost-based prices rather than retail onces minus tariffs. These minutes are then repackaged by the MVNO and bundled with other services, such as voicemail, which the MVNO will provide on its own. For such a business model it is therefore important that there is an appropriate regulatory framework, such as cost-based access prices. In most countries there is no special provision that gives particular status to an MVNO as such. EU regulations provide only that a firm with significant market power must provide access at cost-based prices to 'network operators'. However, the various national regulators have different interpretations of what constitutes a 'necessary condition' to qualify as 'network operator'. The most advanced countries in terms MVNO entry are the UK and Germany. However, even in those countries the subscribers to MVNOs are a very tiny fraction of total mobile telecommunications subscribers. Competition with spectrum licence holders is not so much on voice traffic but rather on value added services. The scope for value added services in 2G mobile telecommunications is relatively modest, because it is mainly based on voice traffic. In the UK, the most successful MVNO was launched under the brand name 'Virgin'. Its main shareholder is the Virgin conglomerate and it shares much of the corporate brand image for entertainment, retailing and travel services. High expectations have been put on the development of 3G mobile telecommunications services, where such multiproduct firms hope to provide value added-based services leveraging the other activities.

---

[32] See, for instance, the case of the UK as described in chapter 2.

[33] Ideally, MVNOs have their own numbers as well as everything else that network-based firms with a spectrum licence have, except for the spectrum and spectrum access equipment such as base stations and base station controllers.

5.8 Price trends in mobile telecommunications

The mobile telecommunications industry had an interesting history of different pricing schemes, which depended on the intensity of competition, product innovation and consumer behaviour.[34] Firms jointly designed prices for services such as connection, monthly rental and usage to steer demand. While the prices for these different services were also used to some extent for fixed line telecommunications, they found much wider and more creative application in mobile telecommunications. This was mainly due to the advent of competition. As fixed line services in most countries were until the 1990s organised within a monopoly framework, there was no need to use different tariff plans as a means of differentiation among rivals. But for the mobile telecommunications industry, with competition came the need for a firm to differentiate itself from its rivals, to segment the market in order to acquire new customers and to relax the intensity of price competition. The evolution of the mobile industry is an interesting illustration of the forces of competition in shaping pricing behaviour.

5.8.1 The early stage: the 1980s

In the early years of the industry (i.e. mainly the 1980s), the typical mobile telecommunications customers were business users who tended to have relatively low demand elasticity to prices and a high level of phone usage in terms of minutes of usage per month. Pricing packages included high connection fees and high monthly rental fees (which, however, frequently included a certain number of free minutes per month of traffic). Usage prices were high, distinguishing between peak and off-peak times. Overall, the price strategies were essentially designed to extract monopoly rents because of the very strong capacity constraints on the radio frequency spectrum: this happened even in markets with two firms, such as the USA and the UK. Table 5.1 refers to the prices in the UK and shows the typical pricing plan of the period for the two firms, Cellnet and Vodafone. Prices remained remarkably constant in nominal terms over the second half of the 1980s, in spite of technological developments that led to general reductions in cost. Other studies confirm that at that time duopoly market structure gave little incentives to lower prices.[35]

---

[34] See also ITC (1999).
[35] See, for instance the studies by Shew (1994) and Parker and Röller (1997) on the USA and Valletti and Cave (1998) on the UK.

Table 5.1 *Retail prices for mobile telecommunications services, UK, 1985 and 1991 (£)*

	Rental per month	Airtime per minute	
		Peak	Off-peak
1985			
Cellnet	25	0.25	0.08–0.15
Vodafone	25	0.25	0.10
1991			
Cellnet	25	0.25	0.10
Vodafone	25	0.25	0.10

Source: BT and Vodafone, quoted in Valletti and Cave (1998).

Table 5.2 *Retail prices for mobile telecommunications services as set by Vodafone, UK, 1993 (£)*

	Rental/month	Peak call/minute	Off-peak call/minute
Businesscall	25	0.25	0.10
Capitalcall	20	0.20–0.50	0.10
Lowcall	12.80	0.43	0.17

Source: Vodafone, quoted in Valletti and Cave (1998).

5.8.2 Initial stages of the non-business market: the early 1990s

This stage was marked by the change from analogue to digital technology and by the entry of a second firm in most of the industrialised countries. The ensuing relaxing of the capacity constraint on the radio frequency spectrum permitted firms to offer different pricing options. This was also the period where offers for non-business users were introduced through alternative pricing plans. To segment the market between business and non-business users lower monthly rentals (and connection fees) were traded off with higher usage fees (and fewer free minutes included in the rental), with incentives to use off-peak times. Table 5.2 illustrates the case of the UK firm Vodafone. The subscription 'Lowcall' was an offer directed towards non-business users. The low rental price was traded off with higher usage prices.

Although prices were starting to fall during this period, they were still considerably out of line with costs. OECD (1996a) shows that operating cost per mobile subscriber were steadily declining and approaching the operating cost per line for fixed telecommunications. However, prices for mobile telecommunications did not follow this trend at a comparable speed, and were in any case still much higher for mobile than for fixed line telecommunications. In the case of mobile telecommunications, fixed charges such as connection and rental fees fell much more rapidly than usage charges. In 1989 the average fixed component of the OECD mobile telecommunications basket was $564 based on PPP, compared to $396 by 1995, a 30 per cent reduction. Over the same period, the average call charge went from $1.35 to $1.10. To accelerate this pricing trend, the OECD (1996b) recommended further entry of firms. The mobile telecommunications sector thus had an opposite trend in pricing than the fixed line firms. In the fixed line sector, prospective market liberalisation called for tariff 'rebalancing'. This essentially involved an increase in fixed charges (in particular rentals) and a lowering of domestic long-distance and international call charges.

5.8.3 Mass market: the later 1990s

This period coincides with the further entry of firms with the granting of DCS (or PCS) licences, thereby further relaxing the capacity constraint. This increased the number of competitors in the market and reduced prices. There was a proliferation of pricing plans, extensive use of handset subsidies to lock-in subscribers and, as a very important innovation, the introduction of pre-paid subscriptions. Pre-paid subscriptions essentially kept the three 'legs' of telecommunications pricing schemes (connection, rental and usage) but modified their meaning. The connection charge was essentially the price of receiving a phone number and a SIM card. The monthly rental was replaced by a flat fee for recharging the account. This charge was not for a fixed period, but could nevertheless be designed as such by setting a maximum recharge amount and a maximum time allowed for recharges. Initially, the maximum time allowed was only a few months; later, it was possible to have over a year. Usage charges were typically higher than with post-paid subscriptions, but users were attracted by the ability to control expenditures and the fact that it was possible to receive calls even after the expiry of credit. In this increasingly competitive context the full range of marketing tools typically used in the consumer goods industry found their application, such as loyalty schemes (e.g. earning bonus airmiles) or incentives for generating traffic.[36]

---

[36] For instance, the Italian firm TIM introduced a scheme whereby the user got 1 free minute for outgoing calls for every 2 minutes of incoming calls generated.

Table 5.3 *Trends in mobile telecommunications pricing, 1992–1998[a]*

	1992	1994	1996	1998	CAGR[b]
Connection	547	410	231	180	−16.9
Monthly subscription	44.9	38.1	34.2	31.3	−5.8
3-minute call	1.04	1.12	0.99	0.95	−1.5

Notes: [a] Figures are unweighted average prices for pre-paid subscriptions in US dollars for the countries in the ITU sample.
[b] CAGR= Compound annual average growth rate (per cent), 1992–8.
Source: ITU (1999).

5.8.4 General pricing trends

Empirical studies show that the price for mobile telecommunications services saw a sharp decline in tariffs in the mid-1990s. Between 1992 and August 1998 the OECD price index for cellular services fell by 29 per cent (OECD, 2000). This figure is probably underestimated the quality adjusted price trend, as it did not take into account service innovations: it did not capture the increased prevalence of flexible packages from 1993 onwards, for instance. In many cases these tariff plans did not greatly reduce the price of mobile services but they did reduce the cost to the user, because they became better suited to different type of usage patterns. Someone making few calls but valuing 'communications mobility' could obtain mobile service for a lower monthly cost than with traditional pricing packages. This development was brought about by competition between firms and by the evolution of cost in the equipment industry (the learning curve and economies of scale). Increased competition due to the entry of new firms, first, reduced profit margins as prices got closer to costs and, second, increased the pressure for further cost reduction via improved efficiency.

A closer look at the price dynamics for the individual service items shows some interesting patterns (see table 5.3). Over the years 1992–8 the average connection charge declined from $547 to $180 (an average annual decline of 16.9 per cent). In comparison, subscription charges declined much less, with an annual decline of 5.8 per cent over the same period; the fall in calling prices was even more moderate, at 1.5 per cent.

The price changes for calls are the result of components that do not all show the same trend. There is not much systematic information available on this, but for some well documented countries (such as the UK) the different components of the overall price can be disentangled. Table 5.4

Table 5.4 *Average revenue, per minute, mobile calls, UK, by destination, 1997–2001[a]*

Destination	1997/8	1998/9	1999/2000	2000/1	CAGR[b]
National (MTF, off-net and on-net)	0.33	0.23	0.18	0.14	−19
of which on-net only	*0.25*	*0.23*	*0.23*	*0.25*	*0*
International	0.95	0.61	0.42	0.45	−17
International 'roaming'	0.96	0.88	0.82	0.76	−6
Average total	0.36	0.26	0.20	0.16	−18

Notes: [a] Figures are annual (£); periods are annual, running from April to March of the following year.
[b] CAGR = Compound annual average growth rate (per cent), 1997–2001.
Source: Oftel data.

Table 5.5 *Mobile call prices, UK, 1999–2001[a]*

	Vodafone	Cellnet	One2One	Orange
1999/2000	0.222	0.206	0.119	0.155
2000/1	0.176	0.165	0.126	0.138
Market share[b]	32	27	18	22

Notes: [a] Figures are the average price per minute (£) of a domestic mobile call, the periods are annual, running from April to March of the following year. The data exclude revenues from short message service (SMS) and interconnection services.
[b] Market shares are based on percentage of subscribers in March 2000.
Source: Oftel data.

shows that from 1997 onward there was a significant decline in the average per-minute price for calls, falling by 18 per cent a year. However, while off-net call prices fell, on-net MTM prices remained virtually constant over 1992–8. International 'roaming' charges also fell comparatively less than the other items.

Another interesting insight from the UK market was the fact that the revenue per minute for each firm in the market was related to the size of the firm. Larger firms tended to have a much higher average price per minute than smaller firms. Table 5.5 shows that Vodafone, with the highest market share, also had the highest price, whereas One2One, which had the lowest market share, also had the lowest price.

Mobile telecommunications services were priced at a premium over fixed voice services. In 1989, the average OECD price per minute of a

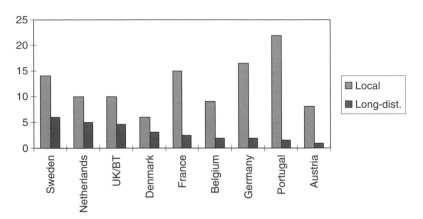

Figure 5.3 *Relative prices of mobile telecommunications services, April 1997*
Note: The figure indicates the multiples by which it is more expensive to call a mobile phone
from the fixed network compared to a fixed local call and a fixed long-distance call.
Source: Author, based on Tarifica data.

mobile call was $0.54: this had fallen to $0.40 in 1998 (a price decline of 3.7
per cent a year, ITC, 1999, p. 81). During the same period fixed line
residential tariffs declined from $0.17 to $0.10 (a 6.1 per cent decline).
This more accelerated price decline for fixed line tariffs was supported in
some countries through price caps, whereas mobile pricing was much less
regulated. However, with mobile telecommunications there was also a
strong quality improvement in the service.

OECD (2000) shows that the mobile's price premium was not based on
cost considerations, as operating costs for mobile telecommunications
were actually lower than for fixed line telecommunications, but that
mobile operators were exploiting the users' willingness to pay for the
additional service element of mobility. Initial high investments and the
capacity constraint of scarce radio frequencies were also used as a justification for high tariffs. In general, national mobile call charges did not take
account of the distance between the calling and called parties. The resulting premium of mobile over local call charges on the fixed telecommunications network could be of a factor of 10 or more, while the premium in case
of long distance calls was not as high.[37] Figure 5.3 shows how many times
more expensive it is to make a call (at peak times) from a mobile phone

[37] Whereas this is the case for most national networks, in particular in Europe, this may not
be the case in large countries such as the USA. Nationwide uniform tariffs were introduced
there only towards the end of the 1990s, starting with the so-called 'digital one rate' (DOR)
plan, introduced first by AT&T in 1998; the other major firms followed shortly afterwards.

rather than from a fixed line phone. The premium over long distance calls is less as on the fixed network prices increase with distance whereas mobile operators tend not to charge for distance. The price premium mobile calls are enjoying is, however, declining steadily. Some mobile firms have started to introduce elements of distance-sensitive pricing, with 'home zone' tariffs approaching the tariffs for fixed line local calls.

These adjustments in relative prices also have led to a redistribution of voice traffic from fixed to mobile services. The market segment most affected by this is domestic long-distance traffic. In spite of substantial price declines, long-distance traffic is declining. Mobile telecommunications traffic in contrast is increasing steadily, which suggests a substantial shift of voice traffic from fixed to mobile.[38]

On balance, mobile operators have the scope to preserve a reasonable price premium over fixed line services as long as any of the following apply:

- Mobile phones give the additional service element of *mobility* for which consumers are willing to pay
- Fixed line operators have no long-run *capacity constraint*, as they can lay as many cables as they need; mobile operators instead face the capacity constraint of scarce radio frequencies.

As more frequencies are allocated to mobile telecommunications and to the extent that mobility is no longer conceived as a special feature of plain voice telephony, price differences between fixed and mobile telecommunications are likely to be reduced further. Reiffen and Ward (1997) investigated for the US market the extent to which the latter may have taken discriminatory actions when providing interconnection services to rival firms. They found that prices increased with the equity share held in the mobile telecommunications firm by the incumbent fixed line firm. There were two sources for this price increase: first, it could simply reflect increased market power as a result of discrimination; second, it could reflect higher efficiency due to better services after mobile and fixed line integration. In that case, in spite of the higher prices, consumer perception of quality would be increased by greater integration. The fact that quantities declined, too, suggests that some of the price increase was due to discrimination.

Mobile telecommunications firms have an interest in exploiting incomplete consumer information about technical features of their services and to enhance this by proliferating pricing schemes. This also increases the

---

[38] As a representative example, take Finland. Long-distance calls declined during the 1990s at an average annual rate of 4 per cent, whereas mobile calls increased at a rate of 55 per cent per year. However fixed lines were more intensively used: they carried on average more than three times more calls and the average duration of a call was twice as long as for a mobile line. See Ministry of Transport and Communications Finland (1998).

difficulties in making fair price comparisons across firms. A detailed investigation (ITU, 1999) showed that pure mobile firms tended to have lower tariffs than incumbent firms that provided both fixed and mobile telecommunications services. The reason for the incumbent's ability to command higher prices resided in the switching cost for users and the reluctance of incumbent operators to engage in aggressive pricing strategies.

5.8.5 FTM call pricing

In all countries, with the exception of those practising the RPP principle, fixed network users have to pay for more for calls to mobile subscribers than for fixed line subscribers. FTM prices are typically the most expensive domestic prices for fixed telecommunications voice services, although prices may vary quite strongly across countries. Table 5.6 shows the prices observed in 1999 for several OECD countries. The cheapest country was Norway, with an FTM price of $0.18 per minute at peak time. The most expensive country was Poland, with an FTM price of $0.69 per minute at peak time. There was thus a huge variance across countries. In general, FTM prices tended to be well above cost. An FTM call at peak time on average in the OECD cost $0.38, which was $0.10 more than the sum of the mobile termination charge ($0.26, a proxy for the origination charge in the fixed network) and the fixed termination charge ($0.02) as determined by the OECD. There was also a huge variation across countries in the reverse direction of the calls, i.e. in MTF prices. The cheapest country was Luxembourg, with an MTF price of $0.10 per minute at peak time. The most expensive country was Germany, with an MTF price of $0.68 per minute at peak time. Table 5.6 also shows that in many cases a FTM call cost much more that a MTF call. At off-peak time a FTM call was, as an OECD average, 33 per cent more expensive than a MTF call.

Mobile operators do not have very strong incentives to lower FTM prices as mobile users do not select their mobile telecommunications supplier according to their FTM tariff. In several countries the market power mobile firms were presumed to have thus induced regulation of FTM tariffs. In 1998, the European Commission initiated an investigation into the pricing of FTM calls, which were deemed as being too high. The Commission derived tariff principles to provide guidance to regulatory authorities in the EU member states, and in countries where formal proceedings where initiated prices typically did fall.[39]

---

[39] In the UK, for instance, as seen in Oftel (2001b).

Table 5.6 *Airtime price of calls between fixed and mobile networks, 1999[a]*

	FTM		MTF		Ratio of MTF/FTM	
	Peak	Off-peak	Peak	Off-peak	Peak	Off-peak
Australia	0.30	0.16	0.31	0.17	1.01	1.06
Austria	0.31	0.20	0.37	0.23	1.18	1.18
Belgium	0.38	0.17	0.32	0.13	0.83	0.75
Canada[b]	0.00	0.00	0.17	0.08		
Czech Republic	0.49	0.49	0.44	0.27	0.90	0.56
Denmark	0.19	0.10	0.22	0.11	1.16	1.17
Finland	0.24	0.15	0.24	0.12	0.99	0.80
France	0.43	0.22	0.22	0.22	0.50	1.00
Germany	0.47	0.24	0.68	0.27	1.44	1.17
Greece	0.47	0.47	0.38	0.38	0.80	0.80
Hungary	0.53	0.35	0.63	0.46	1.18	1.30
Iceland	0.20	0.16	0.18	0.13	0.89	0.77
Ireland	0.26	0.17	0.28	0.14	1.05	0.79
Italy	0.40	0.15	0.27	0.15	0.68	0.98
Japan	0.28	0.28	0.20	0.12	0.71	0.43
Luxembourg	0.35	0.21	0.10	0.10	0.29	0.48
Mexico	0.29	0.29	0.27	0.15	0.91	0.51
Netherlands	0.42	0.24	0.34	0.14	0.82	0.58
New Zealand	0.41	0.41	0.35	0.22	0.84	0.54
Norway	0.18	0.18	0.16	0.15	0.91	0.81
Poland	0.69	0.69	0.53	0.29	0.78	0.42
Portugal	0.50	0.50	0.34	0.34	0.67	0.67
Spain	0.34	0.34	0.62	0.28	1.84	0.83
Sweden	0.35	0.23	0.49	0.17	1.42	0.74
Switzerland	0.38	0.26	0.36	0.24	0.94	0.92
Turkey	0.42	0.42	0.25	0.25	0.60	0.60
UK	0.36	0.24	0.39	0.12	1.11	0.50
USA[b]	0.00	0.00	0.31	0.31		
OECD Average[c]	0.37	0.28	0.34	0.21	0.94	0.78

Notes: [a] Prices are in US dollars converted at PPP exchange rates (February 1999); airtime prices include any set up charge for fixed and mobile networks spread over five minutes, excluding VAT.
[b] Country with RPP.
[c] Average for FTM calls is for countries with CPP.
Source: OECD data.

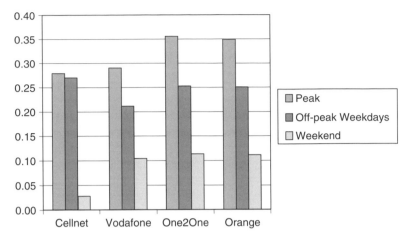

Figure 5.4 *FTM prices, UK (US dollars per minute), 2000*
Source: OECD (2000).

If the telephone number does not reveal whether it is associated with a fixed or mobile line, an issue of *transparency of prices* for the calling party arises. In some countries, such as the USA, the numbering system does not identify the network by the number. But even if this were not so, and as in most countries fixed and mobile lines have different prefixes, the same transparency problem would present itself with the introduction of number portability. 'Number portability' refers to the fact that a customer can retain her mobile number even in the case of a change of service supplier; this helps to reduce customer lock-in and market power of firms. It should also increase competition and lower overall prices to customers.

The UK case shows that the prices for FTM can vary substantially across firms, with new entrants typically having higher termination fees than incumbents. As figure 5.4 shows, in 2000, before number portability was introduced, One2One and Orange had much higher FTM charges at peak time than their competititors Cellnet and Vodafone. However, Cellnet had the highest off-peak prices.[40]

Inspection of a graph of the penetration rate against FTM charges, as indicated in figure 5.5, suggests a negative correlation, although there are significant outliers in this picture. One possible explanation is that a higher level of competition because of entry of firms leads to a lower level of FTM tariffs which again leads to higher penetration rates. This relationship is,

[40] A detailed investigation by Oftel (2002) into underlying costs demonstrated that termination charges were far above cost. The competition authority recommended that all firms be subject to a price cap regulation for termination charges whereby they had to reduce the real prices by about 15 per cent annually over four years.

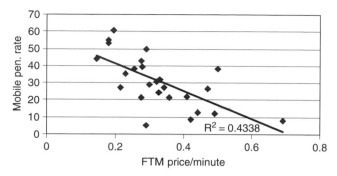

Figure 5.5 *Correlation between penetration rates for mobile telecommunications services and FTM call prices, 1999*
Note: The FTM price is in US dollars per minute, calculated as an average of peak and off-peakprices. The penetration rate is calculated as the number of mobile telecommunications subscribers per 100 inhabitants.
Source: Author, based on OECD data.

however, not particularly strong, as indicated by the R^2 statistic and by the consideration that there are manifold drivers of the diffusion of mobile telecommunications (see chapter 4). For instance, a high FTM price also indicates a high level of subsidy from fixed line to mobile subscribers. Mobile telecommunications firms can use this to subsidise the acquisition of mobile customers, and hence to increase the penetration rate.

The high FTM prices reflect the fact that the interconnection tariffs that mobile firms charge to fixed line firms are high. Table 5.7 shows the interconnection charges in some OECD countries, demonstrating that mobile interconnection prices vary substantially. In countries with CPP, they vary from $0.15 to $0.36 per minute at peak times. In countries with RPP, the rate for calls terminating on the mobile network is typically $0.02–$0.03. This large difference is due to the fact that the receiving mobile customer is paying also for receiving the call. There is also a dramatic difference in the interconnection price of calls in the reverse direction. Typically fixed network firms in a CPP regime pay over ten times more than mobile operators for terminating a call on each other's networks. The OECD average FTM interconnection charge for countries with CPP is more than eleven times the average MTF price. In the USA, with RPP, the FTM and the MTF are exactly the same. Though many regulators concede that termination prices on mobile networks can be higher than for fixed networks, this does not necessarily mean that prices are therefore cost oriented. It becomes increasingly questionable whether mobile operators really have higher termination costs than fixed line operators. OECD (2000) reports several sources which suggest that

Table 5.7 *Interconnection charges between fixed and mobile networks, OECD countries, January 1999*

	FTM	MTF
Austria	22.48	1.94
Canada (RPP)		0.70
Denmark	17.00	1.89
Finland	18.66	1.77
France	32.91	1.92
Germany	35.35	1.40
Italy	28.88	2.96
Japan	29.99	2.28
Mexico	20.00	3.00
Netherlands	34.50	1.69
Norway	15.62	1.80
Sweden	25.59	1.17
Switzerland	29.54	1.93
UK	20.42	0.88
OECD average (excl. RPP countries)	25.97	2.32
United States (RPP)		
Nevada Bell	1.61	1.61
Bell Atlantic	2.64	2.64
Nynex (MA)	3.72	3.72
Nynex (NY)	4.00	4.00
Cincinatti Bell	1.07	1.07
Ameritech	0.83	0.83

Note: The prices are in dollar cents (January 1999) and refer to average interconnect rate per minute for fixed to mobile (FTM) and mobile to fixed (MTF) calls as charged by the incumbent fixed line firm.
Source: OECD data.

average investment per subscriber in mobile networks has fallen below that of fixed line networks. This is equally valid for operating costs. Thus, if termination prices are cost oriented there is little reason for having termination prices on mobile networks higher than for fixed networks.

There are few incentives for these tariffs to rebalance, and in particular to decline. Fixed network firms, particularly when in a dominant position, have a limited interest in reducing FTM tariffs. High FTM prices reduce the traffic competition effect that mobile operators have. Likewise, mobile firms may agree on higher interconnection tariffs because they mean more revenues for them. A high FTM price does not affect their competitiveness, because mobile customers in their decisions to make a mobile subscription do not take into account the FTM termination price. This market failure requires remedy by regulatory measures. The UK regulator Oftel (2002) found that each of the mobile firms had a monopoly over call termination on their network. This is because there was little scope for other technological means in terminating the call differently than on the network to which the called party subscribes. Oftel thus recommended a price cap regulation. Other proposals could involve a switch in the charging regime, either to a RPP or to a bill-and-keep regime. However, neither of these alternatives appears practicable in a CPP regime, and it is thus likely that interconnection prices will remain regulated.

Similar considerations should apply concerning termination charges within an international accounting rate regime. Currently, different rates apply according to whether a call is terminated on a mobile or on a fixed network. Discriminatory pricing may not be justified, and hence regulatory action may be warranted. OECD (2000) reports that the Dutch regulator OPTA did not allow the domestic firm KPN to levy an additional charge when an international call was terminated on a foreign mobile compared to a foreign fixed network in the same country. The motivation was that cost differences were not justified.

The generally prevailing high prices for FTM have created incentives for arbitrage by bypassing these charges. In the case of a large business user, this could be done by transferring FTM calls to MTM traffic through equipment that re-routes traffic from the fixed to the mobile network within the network's own premises. An alternative way could be by channelling traffic via a second country (known as 'tromboning'), using the fact that international termination charges are lower than domestic termination charges for mobile calls.[41] Although the sustainability of these inefficient bypass arrangements may be questioned, they has been applied in several circumstances.

5.8.6 Convergence of mobile with fixed network pricing

The introduction of competition among long-distance fixed voice telecommunications services has consistently reduced prices, but has not made long-distance pricing disappear altogether. Some countries, such as Sweden, have abolished long-distance pricing for fixed calls. Competition from mobile networks gives an incentive for this because within mobile networks national prices are not distance-sensitive.[42] The alignment of

---

[41] In the fixed network similar arbitrage arrangements (known as 'refilling') may occur with international traffic where an operator channels traffic through a third country to circumvent high international accounting rates.

[42] Though there have been attempts to introduce home-zone pricing (for instance, in Germany) these did not have many followers.

Table 5.8 *Airtime prices of fixed telecommunication calls, Denmark, 1999*[a]

	Local		Long-distance	
	Peak	Off-peak	Peak	Off-peak
TeleDanmark	2.5	1.3	4.0	2.0
Mobilix	2.5	1.3	2.5	1.3

Notes: [a] Prices are in dollar cents per minute (July 1999).
Source: OECD data.

long-distance tariffs with local tariffs is more likely to happen as the number of mobile subscribers comes to exceed that of fixed networks. The fact that mobiles have overtaken fixed telecommunications has contributed to the elimination of long-distance fixed pricing in some countries. Table 5.8 illustrates the case of tariffs in Denmark. The mobile network operator Mobilix offered customers of the fixed line operator Tele-Danmark Mobilix fixed line services at uniform national pricing. This new pricing structure cut the price of long-distance calls by 37 per cent and constituted a large step eliminating long-distance pricing in Denmark.

This begs the question why are customers sticking to fixed line operators and why they still charge a price premium on long distance. This can be explained by several facts. Even though outgoing call prices may lead customers to substitute a fixed line connection by a mobile connection, they have to take into account the fact that this will significantly increase the cost of being called, which will increase the cost to users. In terms of service performance, fixed networks can also offer faster data transmission speed which is particularly important for Internet users. Access pricing to the Internet also favours the user of the fixed network: fixed line firms offer discounts on the per-minute pricing to Internet access or 'always-on' connection. This means that the differences between the pricing of fixed network calls and mobile calls are even greater with respect to local calls to the Internet.

The opportunity costs for mobile use are particularly large in a context where local calls are free. OECD (2000) finds that in countries with unmetered local calls rates for fixed networks mobile penetration rates tend to be lower compared to countries where local calls are metered. The difficulty for mobile operators to match such schemes is due to the fact that they have to pay interconnection fees for all off-net calls. The UK mobile firm One2One introduced a service whereby off-peak calls were unmetered, but was forced to discontinue this scheme because of the escalating inter-connection costs.[43]

5.8.7 Effects of different cost sharing regimes

As already seen, from a theoretical point of view the domestic tariff regime (CPP or RPP) should have an impact on pricing, in particular for FTM, with greater price competition under RPP. But regulators do not generally recommend the switch to RPP as a remedy for these distortions. From a policy standpoint, there is still a debate about what is the most appropriate regime. In the end, many practical arguments militate in favour of CPP. The CPP system operates in all OECD countries, except Canada and the USA, where the RPP system prevails (but where nevertheless there are pockets of CPP use). China, including Hong Kong, is in a RPP regime. In Mexico, CPP was introduced on 1 May 1999. Since then, mobile subscribers in Mexico have been able to elect whether they receive such services. The introduction of RPP led to a record subscriber growth in Mexico and to a strong increase in traffic between fixed and mobile networks.[44]

This widespread perception of the growth-inhibiting effect of RPP begs the question why this regime has been adopted at all. The major reason relates to the legacy of countries with unmetered local calls in the fixed network. Whereas it was relatively easy to adopt CPP in countries with metered fixed line local calls, this was seen as more controversial in countries with free local calls. The problem was how to alert the caller whether she was calling a mobile user. In countries with CPP, a numbering system has generally evolved to allocate different prefixes to mobile services as one way to alert users to the fact that they are calling a mobile number and that different charges may apply. This distinction was not needed in countries with RPP.

Table 5.9 summarises the main practical features of RPP and CPP. From this it emerges that in most instances the advantages of one system are the disadvantages of the other. The main advantages of RPP, as perceived by policy makers, are that with RPP, mobile telecommunication pricing is independent of fixed network regulation. Moreover, RPP is a device to apply competitive pressure on fixed to mobile prices as they are internalised by the mobile network. Charges for both incoming and outgoing calls are paid by the same individual who chooses the network operator, and who therefore has the ability to switch to a different firm to obtain better prices not only for outgoing but also for incoming calls. RPP pricing is

---

[43] A frequently quoted anecdote concerning the 'misuse' of this offer was the use of the mobile phone as a 'babyphone'.

[44] See OECD (2000, table 13).

Table 5.9 *Comparison between CPP and RPP*

	CPP	RPP
Budget control	Easy as customer pays only for outgoing calls	More difficult since customer pays for both incoming and outgoing calls
Tariff transparency for FTM calls	Not assured (charges often determined by mobile operator)	Assured (calls from fixed network cost only local call tariff)
Competitive pressure on tariffs	For outgoing mobile call tariffs only	For outgoing and incoming mobile call tariffs
Fixed network regulation	Affects mobile tariffs directly	Does not affect mobile tariffs directly
Bypass	Potential bypass of mobile pricing (e.g. 'tromboning')	No bypass of mobile pricing
User accessibility	User has incentive to keep handset switched on	For cost control reasons, reduced incentive to keep handset switched on

transparent because each firm charges the user only for its network's service. Under CPP, the fixed network operator may charge rates to the fixed user that are determined by the mobile operator; users may not generally realise that the mobile operator determines this price. The numbering allocation is also straightforward, as the call originator does not need to discriminate between numbers to be called. This may be particularly appreciated by businesses, as they can pay for their customers' calls (an '0800' number). From a regulatory point of view there are also strong advantages, as there is less scope for inefficient bypass opportunities such as 'tromboning'.

However, there are also some perceived disadvantages in the RPP system. For instance, it discourages usage. The fact that the receiving party has to pay for incoming calls creates incentives for mobile users to switch off their mobile phone when not placing calls and discourages them from giving out their phone number, with the ultimate effect of reducing the accessibility of mobile subscribers. Mobile users have greater difficulty in fixing their budgets for mobile services and this reduces the scope for pre-paid cards, especially when the rationale of using pre-paid was for receiving calls. The slower development of pre-paid cards may be seen as an obstacle to the diffusion of mobile telecommunications services as they have turned out to be an essential ingredient for rapid market growth under CPP regimes.

As is shown in chapter 4, entry of new firms has provided a strong impetus to market growth. OECD (2000) shows that the introduction of a new firm can in a country be correlated with its advance in the OECD rankings by penetration rates relative to other countries in the year following entry. The two exceptions are Canada and the USA: despite the introduction of new operators, both countries have slipped in their relative OECD ranking. The most likely explanation for this trend is that these two countries, along with Mexico, had RPP. In fact both Canada and the USA started to review their pricing regimes at the end of the 1990s. The FCC released a notice of proposed rule making on 7 July 1999 in which it indicated an intention to remove regulatory obstacles to offering CPP. The FCC said that wider availability of CPP had the potential of developing competition in the local loop and providing an important opportunity for consumers who had not previously used cellular mobile services. This finding was consistent with the experience of OECD countries where CPP was predominant.

The FCC adopted a neutral approach on the tariff regime, arguing that the success of CPP in the USA should reflect a market outcome. User associations, on the other hand, were agnostic about switching to CPP. Some companies were reluctant because there were no standards to alert users that they would have to pay for a call to a mobile subscriber (which cannot be inferred from the number dialled), the risk of bypass and the arrangements for exchanging billing information.

The restraint of RPP on market development therefore appears to be a more recent development. When mobile phones were mainly for business users, the difference between RPP and CPP in terms of subscribers and of traffic was not substantial. Indeed markets with RPP initially even out-performed markets with CPP in terms of subscriber growth. However, as the mobile sector expanded its focus to include personal communications, this ranking reversed.

5.8.8 International 'roaming'

A simplified way of comparing international 'roaming' charges is to compare the cost of international costs made in opposite directions. The OECD (2000) refers to this as the 'call pair' methodology. The price of an international call from a 'home country' mobile network is compared to the price of an international call made from a foreign network by a user 'roaming' in that country. Table 5.10 lists some EU countries by

Table 5.10 *European 'roaming' prices, 1999^a*

Country	Firm	International 'roaming'(1)^b	International call from home country(2)^c	Ratio(1)/(2)
Belgium	Belgacom	1.35	0.98	1.37
UK	Cellnet	1.28	1.35	0.95
Belgium	Mobistar	1.24	1.28	0.97
Netherlands	Libertel	1.20	0.44	2.71
UK	Vodafone	1.19	1.44	0.83
Greece	Panafon	1.15	0.90	1.27
Netherlands	KPN Telecom	1.06	0.30	3.54
Greece	TeleSTET	1.05	0.78	1.34
France	France Telecom	1.04	0.49	2.10
Germany	T Mobil	1.04	1.01	1.03
Denmark	Sonofon	1.00	0.58	1.74
Germany	Mannesmann	0.96	0.83	1.15
Spain	Airtel	0.89	0.64	1.38
Sweden	Netcom	0.87	0.27	3.29
Ireland	Esat Digifone	0.83	0.65	1.28
Portugal	TMN	0.81	0.16	5.03
Ireland	EirCell	0.80	0.72	1.10
Norway	Telenor Mobil	0.79	0.19	4.16
Denmark	TeleDanmark Mobil	0.75	0.55	1.37
Austria	Mobilcom	0.72	0.39	1.85
Finland	Sonera	0.72	0.62	1.16
Sweden	Telia Mobile	0.71	0.47	1.50
Austria	max.mobil	0.67	0.39	1.74
Sweden	Comvik	0.65	0.38	1.73
Finland	Radiolinja	0.62	0.76	0.82
Italy	TIM	0.55	0.50	1.11
EU average		0.92	0.66	1.39

Notes: ^{a} Prices are in US dollars per minute (1999).
^{b} Average international 'roaming' mobile call price from EU area to home country.
^{c} Average international outgoing mobile call price to EU area.
Source: OECD data.

decreasing order of international 'roaming' prices. There is a huge variation. Belgium is the most expensive country, with $1.35 per minute, while the least expensive is Italy, $0.55. There is an even wider variation in prices for international calls, with Norway being the cheapest, with $0.19 and the UK the most expensive, with $1.44. On average, however, 'roaming' is more expensive ($0.92) than making the international call from the home country. Looking at the last row in table 5.10 one can see that the ratio between 'roaming' and international charges is typically greater than 1, which suggests that prices for such services incorporate unwarrantedly high profit margins. With the exception of Finland, the ratio tends to be below 1 in countries with very high international calling charges.

Among the causes of high 'roaming' prices are the absence of competition, absence of regulation, strong disincentives for operators to negotiate lower prices and the fragmented nature of the market reducing the purchasing power of users. Another particular problem is the non-transparency of prices to the user. The complexity of mobile tariff setting makes it difficult, and sometimes impossible, to obtain such information. There are different charges to users depending on which network is being used, the destination of the call and the time of day. In most cases, the choice of the network is made by the handset picking the strongest signal when it is switched on. Given that users do not normally pick the network and are not provided with the information necessary to facilitate the most economical choice it is not clear why they should be asked to pay different prices. It would be much easier for operators to set a single rate for a group of countries and standard times for peak and off-peak calls for all countries. There are technical solutions which could be delivered. For example, least-cost routing could be performed by the handset, using information downloaded from an independent source. A more sophisticated regulatory option is to license MVNOs. Companies could develop brand name services of global or continental scope, encompassing calls to national and international networks, plus 'roaming' services.

Excess profits opens up the questions of regulatory intervention and anti-trust proceedings. The problem for national policy makers is that prices are the result not only of own country competition but depend also on the degree of competition in other countries as well. It is thus not clear whether domestic regulatory intervention alone could improve the situation. This also begs the question whether it is legitimate for national regulators to do so, as any action by national regulators to reduce the wholesale rates charged by national firms would not benefit national consumers. In order for national consumers to benefit from increasing competition at the retail level, national mobile networks must be able to obtain competitive rates from foreign operators. National approaches are not able to address the structure of wholesale charges for international 'roaming' without international cooperation. The European Commission has taken an opportunity to find ways to revise IOT arrangements in a way that encourages competitive offers. It is also suggested that tariff information should be more widely available to users (European Commission, 2001).

5.9 Conclusion

This chapter has analysed the conduct of mobile telecommunications firms with respect to pricing. Pricing in the market is intimately linked with the evolution of market structure. As with many network industries, pricing to a large extent relies on two-part tariff systems. The welfare analysis of pricing plans is complicated by the fact that it concerns competition among networks, and very limited guidance in terms of robust results can be derived from theoretical models. In the early phase of the industry, when capacity constraints were particularly tight, pricing was designed to attract high-usage customers with high fixed prices and relatively low usage fees. As the capacity constraint was relaxed and more firms entered the market, new customer segments with lower-usage profiles were explored. Tariff schemes were based on low fixed prices and relatively high usage fees. The ultimate product with this respect became the pre-paid card.

Price setting in the mobile telecommunications industry is based only slightly on cost and is designed rather to exploit market power. This is most effectively undertaken in the traffic termination services on mobile networks, in particular for traffic originating from fixed line networks. On a mobile network, the firm acts as if it had a monopoly on customer access. The high profits generated from terminating FTM traffic are in many cases used to cross-subsidise other market segments with positive externalities, such as customer acquisition, for the firm. There is a tendency in the market to subsidise the acquisition of subscribers because they can act as attractor for traffic from the fixed line network. A great deal of subscriber and market growth has therefore been made possible thanks to the transfer of termination payments from the fixed to the mobile network. These price–cost deviations have attracted the attention of regulators and they are exploring ways to redress market failures. There is, however, an ongoing debate whether regulatory intervention really can improve on this.

There are alternative charging principles, such as RPP, where such distortions do not arise. However, RPP is in many instances of limited practical value as it has other drawbacks, such as restraining the usage of the mobile network. It is therefore more likely that price regulation will persist in some of the mobile telecommunications market segments where firms can exert market power effectively. The main message from this chapter is therefore that increased competition among networks does not necessary lead to a lower likelihood of market failure due to abuse of market power, and hence the need for regulation may persist.

6   Issues in radio spectrum management

6.1 Introduction

Radio spectrum is a scarce, but indispensable, input in the provision of mobile telecommunications services. Because of the externalities in spectrum usage such as electromagnetic interference and compatibility of equipment, the allocation of spectrum needs to be coordinated both *across* countries and *within* countries. Whereas the international allocation of spectrum is taking place in periodic institutional meetings whose nature has not changed much over three decades, a fundamental change has occurred at the national level where frequencies are assigned to individual firms. The changes mainly concern the method of assignment, with a shift from administrative procedures toward so-called 'market-based mechanisms'. These changes have implications for both the market structure and performance of the mobile telecommunications industry.

Section 6.2 illustrates the international aspects of radio spectrum management, dealing with the allocation of frequencies for different types of wireless communication services. Section 6.3 moves to the national context and describes various methods of assigning radio spectrum to firms. This also includes a theoretical examination of auction theory, illustrating the assumptions under which the most desirable properties of this method actually apply. Section 6.4 describes the spectrum assignment method in practice, by referring to country-specific contexts. Sections 6.5–6.7 describe actual assignment procedures for radio spectrum for 3G mobile telecommunications services in several European countries. Section 6.8 draws some brief conclusions.

6.2 International spectrum allocation

The electromagnetic frequencies suitable for wireless radio transmission are limited to a subset of frequencies of the radio spectrum. The government's

223

task is to regulate the use of the radio spectrum to provide interference-free transmission and reception. Because radio waves do not stop at national borders, spectrum allocations for different users have to be coordinated internationally. In practice, this is done through the ITU, a UN Agency at periodic gatherings (normally every three years) called the World Radio Conference (WRC). The aim of these international coordination efforts is two-fold: first, to decide which frequency blocks should be allocated to which type of service; second, to decide on the technical standards which make the equipment interoperative. Whereas the first aim is intrinsically necessary to provide interference-free communication, the second aim is more peripheral. In the early days of mobile communication, national standards prevailed and so hence features such as international 'roaming' were not available. Only with 2G systems (e.g. GSM) did technical coordination became widespread and international services feasible.[1]

The frequency assignment is a two-stage process: the first stage, referred to as *allocation*, concerns the decision of which and how much spectrum is to be allocated to a particular type of application; the second stage, called *assignment*, concerns the distribution of the frequencies identified to the network operators.[2] The first stage is decided in an international context, the second by national governments. The assignment of frequencies for cellular mobile services has thus been devolved from national administrations to an international body, although the formal assignment of frequencies to the network operators remains with national administrations.

Until the early 1990s (i.e. before digital cellular systems emerged) there was relatively little international coordination in spectrum allocation and technical standardisation. The International Radio Consultative Committee (CCIR) of the ITU recommended basic principles for future analogue systems intended for international use (e.g. international 'roaming') in 1986 (Withers, 1999). In the meantime, however, technological evolution was pointing towards digital systems, such as GSM in Europe. There are in fact at least four different incompatible systems for digital mobile telecommunications deployed (2G systems). They also operate in four different frequency ranges (800 MHz, 900 MHz, 1800 MHz and 1900 MHz).

The WRC responded to the requirements for global, versatile, digital radio communications, in 1992 when it began preparations for the

---

[1] The largest area of international frequency management and technical standards was established by the ITU. To ensure further harmonisation at the European level, the issues are also dealt with by the CEPT. CEPT has handed over the responsibility for frequency issues to the ERC, which also has a permanent body, ERO. As we have seen, ETSI is mainly in charge of harmonisation and standardisation of the equipment to be used. These institutions propose voluntary standards and governments may or may not endorse these standards as mandatory (Bekkers and Smits, 1997).

[2] For more details on these processes, see Bekkers and Smits (1997) and Withers (1999).

development and deployment of relevant systems. This project became known as '3G mobile telecommunications systems' under the heading IMT-2000. The frequency bands in the ranges of 1885–2020 MHz and 2110–2200 MHz were allocated, although in some countries (such as the USA) some of this spectrum was already used for 2G mobile systems. There was also the provision of a satellite element to be integrated with the terrestrial 3G system, for which the frequency bands in the range of 1980–2010 MHz and 2170–2200 MHz were selected. The member countries of the ITU, however, failed to reach agreement, in particular because of the USA's reluctance to free spectrum currently occupied by military applications. The WRC decided in 2000 that individual countries would have an option to choose from three designated frequency bands. However, the USA refused to commit spectrum for IMT-2000 services since they were already using for 2G services the spectrum in the frequency ranges that Europeans would be using for 3G services, whereas the 1800 MHz range, used by Europeans for GSM, was already used by military services in the USA. Relocation of military services to other frequency bands is very costly; the US Department of Commerce has estimated the related relocation costs at $2–5 billion.[3] Thus in the end IMT-2000, if established in all countries, will not be on the same internationally standardised frequency range.

6.3 National spectrum assignment

Radio spectrum turned out to be an economically highly valuable resource whose allocation has strong welfare implications; efficient allocation mechanisms are thus of great public interest. In practice, four mechanisms have been used:[4]

1. First-come-first-served
2. Lottery
3. 'Beauty contest'
4. Auction.

The first three mechanisms can be considered as *administrative procedures* with varying degrees of complexity; the auction is a market-based mechanism. The different mechanisms are now described in more detail.

6.3.1 Administrative assignment mechanisms

First-come-first-served and the lottery are methods by which administrative effort and discretionary choice is essentially limited. Both are therefore rapid and involve little administrative costs. The-first-come-first-served

---

[3] See US DOC (2001).   [4] See Hazlett (1998).

method has been criticised for its allocative inefficiency and the possibility of strategic pre-emption. The lottery is similarly inefficient from the point of view of resource allocation, as the licence receiver may make inefficient use of it. Allocative efficiency may be restored by allowing for resale of the licence, but at the cost of violating the principle of equity.

'Beauty contests', or comparative hearings, are also an administrative attribution process. As a first step, the regulator has to establish the *selection criteria*, also giving an indication of their relative weights. Once the candidates have submitted a proposal, the regulator is called to select the winner(s), consistent with the selection criteria. These proposals may also be complemented by a lump-sum payment for the licence. Depending on the weighting of this payment in the overall scoring system, this method becomes similar to a (sealed bid) auction. The winner is determined as the firm whose business plan has scored best in achieving policy goals.

'Beauty contests' have a series of potential failings. The literature indicates the risk of selecting the wrong candidate, the excessive length of the procedure, inefficient appropriation of scarcity rents, regulatory arbitrariness and insufficient incentive for optimum spectrum usage. The main problems with 'beauty contests' relate to *asymmetric information*. It is difficult for the regulator to assess whether information provided by the candidates is credible. Candidates can back up their claims with evidence of their achievements in the past, but this does not eliminate the problem of credible commitment to future undertakings. Other problems may arise from imperfect regulation (e.g. regulators who are subject to private interest lobbying and therefore biased), or from incompetence. This situation of asymmetric information can be translated into a principal–agent setting, with the regulator being the agent and the government/parliament the principal. This literature has elaborated the risk of 'regulatory capture' – i.e. the fact that the regulator may act in the interest of the regulated party.[5]

6.3.2 Auction

Auctions are typically referred to as a mechanism to assign spectrum to the firm that is prepared to pay most.[6] The purported virtues of an auction in the context of spectrum assignment are a rapid deployment of new services and technologies; recovery for the public of a part, if not all, of scarcity rent the spectrum provides to the winner; and efficient assignment of spectrum to the firm that values it most. However, theory tells us that there are several types of auctions, with their design depending on the rules

---

[5] For a survey of the issues, see Levine (1998).
[6] For related surveys, see Klemperer (1999) and Cramton (2002).

applied. Unless very special conditions are fulfilled, the outcomes from auctions may be very sensitive to these rules. In practical terms the design of the auction is therefore of the utmost importance for achieving any of the above goals.

Auctions may be conducted on a *sealed bid basis* or on an *open outcry basis* in either Dutch format (i.e. a descending-price auction, such as used in the Netherlands to sell tulips) or an English format (i.e. an ascending-price auction). It can be decided whether the winner pays the highest bid (first-price auction) or the second highest bid (second-price auction), or a combination of the bids. Items can be auctioned sequentially or simultaneously, individually or in bundles or, in the case of a combinatorial auction, as a combination of both. From a public policy perspective, a well-designed auction should award the spectrum to the most qualified bidder(s), and these should be deterred from engaging in socially wasteful attempts at market manipulation, while preserving important revenue-generating capabilities. To evaluate which auction format best meets the policy objectives it is necessary to predict bidding behaviour.

Auction theory, which models each auction as a non-cooperative game, can provide some useful guidelines. The equilibrium outcomes of the game are used for recommendations concerning the most appropriate game format. The most important paradigm with this respect is based on Vickrey's (1961) 'revenue equivalence theorem', which claims that, given a certain set of assumptions, all of the above auction formats are efficient in the sense that the item being auctioned is awarded to the bidder who values it the most. All formats will also yield the same expected revenue. The four necessary assumptions for the 'revenue equivalence theorem' to hold are:

1. *Independent private value*: Each bidder i knows v_i, the value of the item under auction to herself, but is uncertain about its value to other bidders. The seller is uncertain about its value to the bidders. Each bidder cares about other bidders' valuations only insofar as it affects her bid.
2. *Symmetry*: The seller and the other bidders believe that the value of bidder i is a random variable drawn from a known distribution function that is common to all bidders.
3. *Normalisation*: The payoff to a bidder who loses is zero.
4. *Risk neutrality*: If bidder i wins the auction paying m, her payoff is $v_i - m$. The assumption that payoffs are measured in a currency implies that bidders are risk-neutral.

Several of these assumptions may not be realistic in the case of spectrum auctions. For instance, it may be very hard to know in advance the value of the licence, since this will depend on the evolution of demand, technology and competitors' responses. Moreover, valuations are likely to be correlated among bidders (e.g. $v_i = v + e_i$), as they may have similar expectations about the underlying driving variables. This may also explain the possibility of the 'winner's curse'. Winning a first-price sealed bid auction may not be necessarily good news, as it may indicate that the winner has overestimated the value of the licence. The source of the 'winner's curse' resides in the fact that while each value estimate is unbiased, the highest-value estimate is necessarily biased upward. The revenue equivalence theorem therefore no longer holds. Milgrom and Weber (1982) have shown that in such a context the seller achieves the highest price from selling a single object in an English auction, followed by the second-price sealed bid auction; the Dutch auction generates the lowest expected revenues. In the case of asymmetric bidders – i.e. bidders with different information about the value of the licence – only the English auction remains efficient (Chakravorti et al., 1995).

When multiple objects are auctioned, it turns out that simultaneous auctioning is more efficient than sequential auctioning. With sequential auction, a decreasing sequence of prices can be observed, although the auctioned objects may be identical. This is known as 'afternoon effect' among wine dealers (McAfee and Vincent, 1983), and can be explained by the willingness of risk-averse bidders to pay a risk premium at an early stage. Risk aversion also implies that prices are higher in a first-price auction than in a second-price auction. The reason is that submitting one's true valuation remains a dominant strategy in the second-price auction, whereas risk-averse bidders are willing to pay more than risk-neutral bidders to avoid the loss from failing to win the object.

Klemperer (2002) argues that ascending-price auctions are particularly vulnerable to collusion and likely to deter entry. Sealed-bid auctions are therefore better in this respect, but have two other drawbacks: first, an inefficient firm may win; second, a 'winner's curse' may occur. Because in the real world conditions are rarely those requested by theory, the auction has to be designed very carefully, avoiding all pitfalls, for it to deliver an efficient allocation of the object. Binmore and Klemperer (2002) claim that economists have been advocating the auctioning of spectrum since the late 1950s, following the pioneering contribution by Coase (1959). But it took more than thirty years of persuasion of policy makers to see the first auctions in practice in New Zealand in 1990 and in the USA in 1994.

Two considerations made the recourse to market-based mechanisms appealing for the assignment of licences. The first consideration is efficiency. Market forces were considered the best means for allocating the spectrum to the firm that valued it most and which therefore should be able to use it most efficiently. The profit maximising incentive was thus thought to coincide with the public interest. The second consideration was administrative simplicity. This was particularly valuable when there were a huge number of licences to be allocated (such as the several hundreds of A and B block PCS licences to be assigned in the USA). There was also the fear that extensive litigation could derive from administrative allocation mechanisms, and that this could be avoided with non-discriminatory market mechanisms such as auctions.

6.3.3 Comparison auction vs. 'beauty contest'

Since the introduction of 2G mobile telecommunications systems the practical methods of spectrum assignment have been reduced to two: the 'beauty contest' and the auction. There are several benchmarks by which the two attribution methods can be compared:

1. *Transparency*: The auction method provides much greater transparency than other forms of administrative allocation. Once the auction rules are established, there is much less scope for regulatory arbitrariness.

2. *Collusion*: It may be useful to distinguish two levels at which collusion can occur. One level is at the assignment (pre-entry) stage and one at the post-entry stage. Collusion such as coordination of bids at the pre-entry stage may be particularly damaging for an auction, as it invalidates the main objectives – identify the best candidates and raise licence fees. It is therefore necessary to create competition for the licence. Secrecy in administrative procedures is essential. Collusion is facilitated in multiple-round auctions, as they provide scope for signalling and bid coordination. The auction design therefore has to provide particular provisions to limit the risk of collusion during the auction process. There is nevertheless the risk that escalating licence fees may be used as a credible commitment to practise post-entry collusion.

3. *Discrimination*: Auctions have the advantage that any kind of discrimination has to be made explicit at the outset. If the regulator wishes to favour small firms, this can be catered for by including a mark-up on the bids offered by such firms. (For instance, a bid of 100 is increased by 10 per cent, and the small firm bidding 100 would win the licence as long as the other bidders bid less than 110.) Such positive discrimination is much more controversial in a 'beauty contest', as it would be much more amenable to regulatory arbitrariness.

4. *Cost and speed of procedure*: A well-designed auction is generally less costly to implement and much faster to conclude than a 'beauty contest'.

6.3.4 Caveats for auctions

Although at a theoretical level auctions may appear superior to administrative procedures, there are however also some drawbacks with an auction. Demand reduction and collusive bidding are the two main issues that limit both the efficiency and revenue-generation capacity of multiple-item auctions. With respect to demand reduction, consider the following. In an ascending-price auction for a single item, the dominant strategy is to bid up to the private valuation. If there are multiple items, then the bidder may have incentives to stop bidding for the second item before the private valuation has been reached.[7] Otherwise, continuing to bid for two items raises the price paid for the first. A bidder with the highest value for the second bid may in fact be outbid by another bidder demanding just a single unit.

With respect to collusive bidding, the simultaneous ascending-price auction can be seen as a negotiation among bidders, where bidders can use the bids ('code bids') to communicate to each other on how to split up the licences. 'Retaliating bids' can be used to sanction deviants and thus enforce signalled splitting proposals. The auction closes when bidders agree on a split.[8] To reduce the impact of these strategies, some remedies can be adopted. Bid signalling can be made less effective through concealing the bidders' identities because there is then less scope for retaliating bids. Mechanisms such as 'click-box bidding', where the bidder indicates in a click box the bid increments, also reduces possibilities for code bidding. To lower the incentives for demand reduction, sufficiently high reserve prices can be set. Higher reserve prices also restrict the number of rounds that can be used for coordinating the splitting. Finally, for identical items, inefficiencies for demand reduction can be reduced through a Vickrey auction (Ausubel and Cramton, 1996).

On most scores, the auction appears to perform better, and overall efficiency reasons make this the increasingly privileged mechanism. However, as Melody (2001) pointed out, this may not ensure overall efficiency. As mentioned before, the assignment of radio spectrum can be seen as a two-stage process: in the first stage the spectrum for the type of service is allocated; in the second stage licences are assigned to individual network operators. If market forces in the assignment procedure are adopted only in the second stage, price distortions may occur. If too small a block of

---

[7] See Ausubel and Cramton (1996), who show that with multi-unit uniform-price auctions every equilibrium is inefficient. In presence of bidders that want to bid for more than one unit, there is an incentive to reduce bids and this incentive increases with the quantity being demanded.

[8] See Cramton and Schwartz (2000) for a description of this strategy in the context of US spectrum auctions.

spectrum is allocated to rapidly growing services, artificial scarcity is created and the market mechanism in the assignment of individual spectrum licences is likely to lead to artificially high prices. It may also be difficult to introduce a market process at the allocation stage, but nevertheless the foundation of the efficiency of spectrum management is the allocation process. If this process is inefficient, no assignment method can compensate for it.

The spectrum assignment method can also introduce inefficiencies of its own into the process. The problem is to find the optimum level and structure of the licence fee. Spectrum auctions represent an approach to determine licence fees that relies on the particular conditions in a spectrum market or submarket. Many economists favour auctions because they rely on market outcomes rather than judgements about frequency assignments. However, to achieve maximum spectrum efficiency through an auction, the administrative allocation of the spectrum for the particular uses in question needs to be optimal and the bidding markets perfectly competitive. If either of these conditions does not hold, the fees derived through an auction could be far from optimal. Any inefficiencies in the allocation of spectrum will be carried forward to the auction process and compounded. Any artificial scarcity due to the allocation of spectrum will result in monopoly or oligopoly spectrum prices, and not in competitive market prices. This will distort both the spectrum resource market and the spectrum-using services market by artificially restricting competition in both.

In certain circumstances spectrum auctions may by justified, even if they lead to monopoly or oligopoly outcomes. This is not on grounds of improving efficiency but rather on grounds of equity and distribution. Auctions may become a means of creaming off rents due to resource scarcity, but effective competition cannot be promoted in the spectrum applications market.

These issues have become relevant in the context of the escalating licence fees that have occurred in some European countries for UMTS licences, as will be seen later. The root of the problem of escalating licence fees may be sought in inadequate allocation of spectrum for 3G mobile services. At the WRC in 1992, the spectrum allocation for 2G cellular telecommunications services was decided, based on the anticipated growth in demand at that time. Only a tiny part of the spectrum (a total of 230 MHz) was reserved for future 3G services. Within this framework, national governments assigned the spectrum based on the needs and priorities of individual countries. For instance, mobile telecommunications experienced an unpredicted market success and hence pressure mounted on governments to assign additional spectrum for mobile services, in spite of the fact that technological innovation and the switch to digital technology permitted the use of the existing spectrum much more efficiently. The issue of allocating spectrum for 3G services was left to the WRC in 2000. At that conference, there was, however, little pressure to increase the spectrum for 3G services (519 MHz were allocated). The outcome of the 2000 conference was basically that the spectrum block used for 2G technologies could also be used for 3G, and spectrum capacity could be extended to 3G. Melody (2001) reports that countries which had used the auction mechanism were reticent about extending spectrum capacity. It therefore seems that the 3G auction occurred in a context of artificial scarcity. Proposals to avoid this problem involve eliminating the two-stage setting of spectrum management with a spectrum trading regime where spectrum is assigned without tying it to a particular technology.[9] There is already a precedent of this kind for New Zealand, with mixed results, and the institutional practicability of such a solution is debatable.

6.4 Spectrum assignment in practice

6.4.1 Administrative methods

In most countries, the incumbent fixed telecommunications operator initially operated mobile telecommunications services under a monopoly regime. Some other countries such as the USA, Sweden and the UK instead selected to set up the cellular mobile telecommunications industry in a duopoly framework. In Sweden and the UK, nationwide licences were awarded. One licence was automatically assigned to the fixed network operator, while the second was allocated through a 'beauty contest'. As long as nationwide licences were granted, administrative methods were valuable, but this changed when the number of licences became very large. The USA did not assign nationwide licences, the country was divided into 305 MSAs, defined by a region of at least 100 000 inhabitants, including a town of at least 50 000 inhabitants. The FCC also defined 428 RSAs, with an average population of 150 000.[10] For each MSA, two licences were assigned through a 'beauty contest'. One was earmarked for the local fixed line operator (the so-called 'wireline licence') and one attributed to new entrants ('the non-wireline licence'). The response to the invitation for bidding was unexpectedly strong. It should be pointed out that there was not just one 'wireline company' in each MSA, so there was also effective competition for the wireline licence, with generally more than one bidder. The main selection criteria were the extensiveness of the area that an applicant proposed to serve and the efficiency with which it would use the spectrum assigned. Service prices were also taken into account, but only as a minor consideration.

---

[9] See, for instance, Valletti (2001).   [10] For a detailed description, see chapter 3.

The FCC became administratively overburdened in assigning all the licences in one step, so selection had to be made in stages. The thirty largest MSAs were dealt with in round one; further rounds were organised at five-month intervals. In the successive rounds, the number of bidders increased significantly. For instance, in round three (concerning the MSAs ranking from sixty-one to ninety in terms of inhabitants) there were more than sixteen applications per licence. The FCC could not cope with processing such a large number of bids and so the licences from round three onwards were awarded by lottery. The numbers of participants increased as the pre-qualification criteria for taking part in the lottery were relaxed. It also happened that winners of licences did not always have the resources for implementing a mobile telecommunications network, and in 1987 the FCC determined that it was legal to trade the licence. The lottery for the thirty licences in round four attracted 5182 applications, a number bound to increase in the future. Subsequent lotteries were organised in six rounds between February and May 1986; there were 92 000 applications for the remaining 185 licences. For the RSAs, the FCC adopted a lottery for assigning licences but to avoid opportunistic behaviour asked for financial guarantees and a more rapid deployment of the network, reducing the implementation lag from three years to eighteen months.

Overall, it took the USA four years to award cellular licences for its MSAs alone and seven years including the RSAs. The switch to the lottery mechanism as a way of accelerating the procedures thus can be considered only a limited success. Shew (1994) reports that entering a lottery implied a cost for the bidder in the range of $250–$5000, which many bidders considered limited compared to the expected gains from the cellular market. The sheer number of applicants, averaging almost 500 per licence, slowed the assignation process. The fact that some lotteries were won by firms unwilling or unable to efficiently operate a cellular network led to welfare costs estimated at up to $1 billion (Hazlett and Michaels, 1993).

There were asymmetries in the opportunities for bidders. In general, it was far easier for the wireline operators to deploy their networks because they were more likely to win a licence and could therefore spend more efforts in planning ahead. This emerged clearly from the network deployment figures. By the end of 1984, twenty-five networks allocated to wireline licences were in operation with only nine from non-wireline licences. This gap increased in 1985, with eighty wireline licences and only fifteen non-wireline licences active. To ease entry for non-wireline licencees, they were allowed to 'roam' on the networks owned by wireline licence holders until their own network was operational. By the end of 1986, most of the ninety largest MSAs had two competing systems.

In the top ninety MSAs there were some eighty different licence holders according to the original awards. By 1992 the top twelve cellular firms, of which the largest was McCaw Cellular, had accumulated ninety-one licences for areas with a total residential population of 65 million, serving nearly 60 per cent of the US population. This consolidation was the due to the fact that wireline firms acquired a large number of non-wireline licences from the original licence winners. There also were entrepreneurs such as McCaw who were convinced of the long-term business potential of cellular mobile telecommunications licences. This led to an increase in the value of licences: calculated on a per-inhabitant basis, they rose from $8 in 1984 to $270 in 1990 (Garrard, 1998).

6.4.2 Early auctions

New Zealand

On a worldwide basis, the first spectrum auction for cellular services took place in New Zealand in 1989. This happened in the context of a radical programme of deregulation in the telecommunications sector. The liberalisation attempt was quite far-reaching, as it allowed the winners to trade spectrum or to change, in principle, the use for which spectrum had originally been acquired. The spectrum initially allocated to cellular mobile telecommunications services was on several frequency bands that were tendered during the first half of the 1990s.[11] The results of this auction for cellular services are listed in table 6.1. For the cellular spectrum a sealed bid, second-price (Vickrey) auction was used in 1990. The winning bidder had to pay the price of the second-largest bid. The firms had to bid for frequency blocks linked to particular technologies: one block for AMPS and two blocks for TACS technology. At that time, the incumbent telecommunications firm Telecommunications New Zealand was already operating an AMPS system. The main problem in the auction was that there were not many bidders. Two firms only, namely Telecom New Zealand and Bell South, undertook the bidding for three licences. Thus only two frequency blocks of 15 MHz each were assigned for a total of NZ$36 million. On top of this, Telecom New Zealand had to pay NZ$6 million as an incumbency right for its existing AMPS frequency block (20 MHz). One TACS frequency block was not assigned.[12] To avoid conflict with the competition legislation which mandated three firms, one band had to be re-auctioned in 1993. Because the Vickrey auction was apparently poorly understood, it was in

[11] See Ministry of Commerce of New Zealand (1995). For a comparative description with US auctions, see Crandall (1998).
[12] This seemed to vindicate the recommendation made by government consultants (NERA, 1988), which considered the market viable for only two firms.

Table 6.1 *Radio spectrum allocation, New Zealand, 1990–1993*

Date	Auction format	Number of licences	Total frequency	Total revenue (NZ$ million)
1990	Sealed bid, second-price (Vickrey)	2	30 MHz	36.37
	Incumbency payment	1	20 MHz	6.00
1993	Sealed bid, first-price	1	15 MHz	13.60

Source: Ministry of Economic Development of New Zealand data.

the meantime decided to change the auction format (Ministry of Commerce of New Zealand, 1995). The re-auction of the TACS frequency block was made in a first-price sealed-bid auction. This auction was eventually won by Telstra, paying NZ$13 million. However, Telstra never started to build the network and eventually sold the spectrum to Bell South in 1997, at a price that was believed to be less than 20 per cent of the original fee paid. The government since then has been reconsidering the auction design. To improve the feedback from other participants on the auction of an asset of uncertain value and to avoid the risk of the winner's curse, the system was modified. Subsequent auctions became an Internet-based, multiple-round, ascending-bid auction.

The USA

As seen in chapter 3, the assignment method of the cellular mobile telecommunications licences proved to be very difficult in the US context. For the provision of additional spectrum to the analogue mobile telecommunications services the FCC made several changes in the spectrum assignment procedures. The FCC changed the allocation method, opting for auctions, mainly with the idea of speeding up the process. In 1993, the US Congress approved legislation that allowed the FCC to do this, this new legislation was soon applied for the auctioning of the so-called 'narrowband' and 'broadband' PCS licences. The broadband PCS licences were for the provision of mobile telecommunications services. Because the introduction of this legislation was controversial, substantial delays took place. In the end, it took six years until the first PCS licences were granted in 1995. Regulatory delays were again the main culprit for the late introduction of new services. Rohlfs, Jackson and Kelley (1991) have estimated that the delay in providing additional sprectrum created welfare cost of $80 billion (i.e. about 2 per cent of US GDP).

There was also a change in the definition of the franchises. Instead of using the same MSA and RSA definitions adopted for the original AMPS licences, the FCC proposed using fifty-two larger units known as MTA, subdivided into 493 BTA. For each MTA there would be two licences (A and B licences) with 2×15 MHz each and one 2×15 MHz licence for each BTA. Each BTA would also have three licences (D, E, F licences) with a smaller frequency band of 2×5 MHz. Overall, any town in the USA should in principle be covered by up to six PCS licences. MTA licensees were required to cover 67 per cent of the population after ten years, and BTA licences obliged firms to cover 25 per cent of the population after five years (see table 6.2).

The MTA auctions (A and B blocks) started in December 1994. The licences were assigned using a simultaneous multiple-round auction. This was similar to a traditional ascending-bid 'English' auction except that, rather than selling each licence in a sequence, large sets of related licences were auctioned simultaneously. The FCC's aim was not so much to raise revenues but rather to ensure efficiency in the sense of assigning the licences in a timely manner to those firms best able to make use of them (FCC, 1997). The auction lasted until March 1995: with ninety-eight days for 112 rounds of bidding, it was the fastest among the PCS auctions. The auctions yielded a total of $7.8 billion in licence fees, which was considered a large sum. Nevertheless the suspicion of collusion among bidders to keep fees low arose, as several phenomena consistent with collusive behaviour were detected. Simultaneous and open auctions in theory facilitate collusion: bidders can observe each other's bids and credibly coordinate a collusive agreement by enforcing it through a punishment strategy in case of deviation. For instance, bidders can recur to 'code bidding' which consists in indicating the market number in the trailing digits of the bids for a licence on which a bidder is keen. Alternatively bidders can use an unlimited number of withdrawals to emphasise punishment bids, which consists in bidding up the price on a licence on which the firm is actually not interested.[13] The market structure emerging from these auctions was much more concentrated right from these beginning than with previous allocation methods. Large firms conquered most of the licences: incumbent firms, along with Sprint's mobile subsidiary WirelessCo, won twenty-nine out of the ninety-nine available licences (see table 6.3). Overall, the first three firms received almost two-thirds of the total licences granted.

The rate of licence sales in the secondary market can be considered as a measure of efficiency in the assignment mechanism. A high resale rate would indicate that the initial allocation was relatively inefficient. In this

[13] See Cramton and Schwartz (2000) for a detailed description of these strategies.

Table 6.2 *US Broadband PCS auction results*

Blocks	Total spectrum (MHz)	Service areas	Number of licences	Auction period	Number of rounds	Total auction revenues (US$ million)	Auction revenues per person per MHz ($)
A, B	60	MTA	102[a]	05.12.1994–13.03.1995	112	7,721	0.52
C	30	BTA	493	18.12.1995–06.05.1996	184	9,198	1.33
C (re-auction)	30	BTA	18	03.07.1996–16.07.1996	25	905	
D, E, F	30	BTA	1479	26.08.1996–14.01.1997	276	2,517	0.33

Note: [a]Including the three pioneer preference licences.
Source: FCC data.

Table 6.3 *MTA (A and B block) licences, by firm, 1995*

Firm	Number of MTA licences	Population covered (million)	Winning bids ($ million)	Winning bid/population ($)
WirelessCo	29	145	2,110	15.24
AT&T	21	107	1,686	15.75
PrimeCo	11	57	1,107	17.67
American Portable	8	26	289	11.12
Other	28	117	251	2.14
Total	99	452	7,019	15.54

Source: Author using FCC data quoted in Cramton (1997).

respect, the auction does not compare particularly well. Twelve out of the ninety-nine A and B block licences were resold in 1996; in comparison, in 1991, there were seventy-five resales of about 1400 cellular telephone licences distributed by lottery from 1984 to 1989 (Hazlett and Michaels, 1993). This suggests that there is room for improving the auction design. Several studies have investigated strategic bidding by participants in the MTA auctions. Ausubel *et al.* (1997) estimated benchmark regressions for the determinants of final auction prices, including variables reflecting the extent to which bidders ultimately won or already owned a licence in the adjacent area. They found significant evidence of synergies, with higher bidding prices when the highest losing bidders had adjacent licences. Moreton and Spiller (1998) found similar evidence that combinations of MTA licences were worth more than separate ones: on average, having an adjacent licence added about 25 per cent of value to an individual licence. For the C block auctions the effect was smaller, adding on average some 9 per cent of the licence's value.

The C block licences were originally intended as an encouragement to local entrepreneurs. As each individual BTA would cover on average one-tenth of the population of a MTA, it would be much harder for any firm to obtain wide areas of coverage. Several 'affirmative action' provisions were originally made for the licensing process: for instance, 'designated entities' (such as small firms owned by women or minorities) would qualify for a 25 per cent discount on the bid price and could spread payments over six years. After a legal challenge followed by a Supreme Court ruling[14] the references to minorities and women eventually had to be dropped and only

---

[14] This is known as the *Adarand* Case. For a description, see Hazlett and Boliek (1999).

the preferential treatment for small firms was upheld. As a result of this legal battle, the original auction design had to be reassessed, delaying the auction for nine months. The auction started in December 1995 and ended in May 1996. After 184 rounds of bidding a total licence fee of $9.2 billion was raised. This means that the price of a licence, if calculated on the basis of per-inhabitant of the area the licence refers to, was much higher than for the MTA licences (see table 6.2). This price differential is surprising. Game theory predicts that with items sold in sequence the price declines for items sold later on. The reasons for this are risk-averse bidders and market advantages associated with early entry. Designated entities were granted exceptional support, such as instalment payments instead of up-front payments, as well as subsidised loans. The instalment subsidy was substantial, with the subsidy equivalent to about 28 per cent of the net bid (Hazlett and Boliek, 1999). Nevertheless, the instalment payments and the interest subsidy together accounted for one-third to two-thirds of the increase in the bids of the C-block auctions (Ayres and Cramton, 1996). Hazlett and Boliek (1999) explain the residual through opportunistic bidding – i.e. the adoption of a bidding strategy based on revenue projections that are not fully justified assuming compliance with the terms of the financing contract. The incentives for opportunistic bidding are due to the low interest and deferred instalment financing terms given to designated entities. This shifts the downside risk from the bidder to the government, since the bidder can default on its promised payment to the government if the licence value falls below the net auction price.[15]

This escalation of licence fees as a result of the auction process also led to the first firm failures. The firm BDPCS failed to pay for the seventeen licences it had won, as did National Telecom for its own licence. BDPCS had an aggressive bidding strategy, speculating on the fact that winning a licence would attract the backing from a large firm without a licence. However, this possibility did not materialise and the eighteen licences were re-auctioned in July 1996. In the end, 493 licences were distributed to some eighty companies. The market structure was quite fragmented, with the first three firms, as we have seen, accounting for one-third of the total licences granted.[16]

The auction for the remaining 1479 narrowband licences (D, E, F block) started in August 1996 and ended in January 1997. Any firm

---

[15] Such considerations became relevant when assessing the bids for UMTS licences in Europe in 2000. The financial market 'bubble' led to very lax financial constraints for firms.
[16] NextWave's total value of the bids was $4.7 billion. The firm, however, had problems in paying the further instalments of the licence fee after the first down payment. Long legal battles ensued, with the FCC trying to re-auction the spectrum. In January 2003 the Supreme Court finally ruled that the FCC's decision to cancel the licence for NextWave was wrong (see *New York Times*, 28 January, 2003).

could bid in these auctions, including existing mobile telecommunications operators. Again, the F block was reserved for designated entities. The auction raised a total of $2.5 billion, which was less than the A, B and C block if calculated per MHz. This reflected the perceived lower utility of so called 'narrowband' services, but also collusive behaviour. Cramton and Schwartz (2000) investigated the scope of collusion in the simultaneous open ascending-price auction method adopted. Bid signalling behaviour was particularly prominent in the case of the D, E, F auctions.[17] Although the fully transparent design provided bidders with a great deal of information and eased the efficient allocation of the licence, it also gave incentives for collusive behaviour. The transparency facilitated arbitrage across substitute licences, promoting the efficient agglomeration of complementary licences. It was found that a small fraction of bidders (six out of 156) frequently used collusive behaviour and were sometimes successful in keeping prices low. These bidders won 476 out of the 1479 licences on sale, for some 40 per cent of the population coverage. These bidders also paid significantly less for their licences.

6.4.3 GSM licensing in Europe

With the advent of digital mobile telecommunications technology, European governments started to refine their licence assignment procedures. Whereas the incumbent analogue mobile telecommunications firm had typically been assigned the first GSM licence,[18] the second GSM licence was assigned through a tendering process. This was typically an administrative method, involving the evaluation of sealed bids. The bidding documents normally contained a description of the network build-out and in several cases also an indication of the cash payment to the government for the spectrum use. (In that case, the procedure resembled a sealed bid auction.) Table 6.4 shows in inverse chronological order the licence fees paid in several EU countries, which can account for up to 50 per cent of the initially planned network investment cost. The prices paid are high, not only in absolute terms, but also if related to indicators such as population. In Austria, the licence fee per head is as high as $50 per inhabitant. In some countries, to preserve fair competition among operators, these licence fees were required to be matched by the incumbent firm as well

---

[17] Since bidding amounts in dollars are typically six digits or more, bidders can use the last few digits of a bid to encode messages. See Cramton and Schwartz (2000) for a description of the practices and the interpretation of 'strange bids'.

[18] The exception is Greece as there was no analogue mobile telecommunications firm. The fixed line incumbent monopolist, however, did not manage to receive one of the two GSM licences; the firm was awarded a GSM 1800 licence at a later stage.

Table 6.4 *Licence fees for GSM 900 MHz spectrum*

Country	Licence fee ($ million)	Fee/head ($)
Austria	397	50
Belgium	294	29
Spain	648	17
Greece	164	16
Italy	460	8
Ireland	24	7

Source: Author, based on press releases by firms.

(e.g. Austria and Belgium), or compensation schemes were set up (e.g. Italy and Spain).[19] The question arises as to the limit at which spectrum fees become too high and constitute artificial entry barriers that drive operators from the market, and may even slow down market development. In the USA this happened, as with the spectrum auctions in 1997 there were some companies that went bankrupt because they were unable to find sufficient funding actually to pay the high licence fee offered in the first place.[20]

In cases where assigning the second licence also involved a significant licence fee, the incumbent had to match the winning bid or provide other forms of compensation to avoid charges of unfair competition.[21]

The EU *Mobile Telecommunications Directive* of 1996 established that there should not only be at least two different suppliers of GSM services in the 900 MHz frequency band, but also at least one DCS 1800 operator. Thus by the end of 1998, in almost all countries, there were at least three operators for mobile telecommunications services. In most countries where more than two firms received a licence, licencing was sequential (i.e. first the third operator was chosen, then the fourth). Table 6.5 lists the entry dates for firms in the different countries.

For GSM 900 firms, the regulatory environment and the risk of additional competitors appeared to quite predictable. They were bidding for a licence in a duopoly context; any additional entry at the GSM 1800 level was perceived as being not directly in competition, at least initially, because

[19] In those countries, the competitive tendering was only for the second GSM licence whereas the first GSM licence had been granted automatically to the mobile subsidiary of the incumbent fixed line operator.

[20] This mounted pressure on the FCC to renegotiate and eventually to reopen the auction (see *Financial Times*, 2 September 1997).

[21] The European Commission ruled in this sense. To preserve fair competition among operators, licence fees in Austria and Belgium were matched by the first operator as well, while in Italy and Spain compensation schemes were set up.

Table 6.5 *Entry dates for GSM firms, EU countries, 1992–2000[a]*

	Number of GSM firms	GSM 900	GSM 1800
Austria	4	12.1993; **10.1996**	**10.1998; 5.2000**
Belgium	3	1.1994; **8.1996**	**3.1999**
Denmark	4	7.1992; 7.1992	1.1998; 3.1998
Finland	4	1.2000	3.1998
France	3	12.1992	**5.1996**
Germany	4	6.1992	**5.1994; 10.1998**
Greece	3	**7.1993**	1.1998
Ireland	2	**3.1993**	3.1997
Italy	4[b]	10.1992; **10.1995**	**3.1999**; 5.2000
Luxembourg	2	7.1993	4.1998
Netherlands	5	7.1994; 10.1995	10.1998; 1.1999; 2.1999
Portugal	3	10.1992; 10.1992	**9.1998**
Spain	3	7.1995; **10.1995**	**1.1999**
Sweden	4	9.1992; 9.1992; 11.1992	1.1998
UK	4	7.1992; 1.1994	9.1993; 4.1994

Notes: [a] The dates indicate the date of first supply of services, dates in **bold** indicate a licence obtained in a sealed-bid auction.
[b] The number of firms has been reduced to three, as the latest entrant declared itself bankrupt and its spectrum was equally divided among the three remaining firms.
Source: Author, based on information from European Commission, ITU and Mobile Communications.

the customer equipment was incompatible. But before too long, as 'dual band' handsets were developed, it became clear that the incompatibility had disappeared and services in both the 900 MHz and 1800 MHz frequency range became in principle interchangeable. At the same time GSM 900 firms began to have capacity problems because the frequency range allocated to them could not accommodate the traffic from the rapidly increasing user base. In most countries, GSM 900 firms eventually received frequencies in the 1800 MHz range, too. With dual handsets, GSM 900 operators could overcome the capacity problem they were facing in the 900 MHz frequencies and therefore strengthen their market position. Two types of operators may be envisaged in the long term: pure DCS 1800 operators on the one hand and dual GSM900/DCS 1800 operators on the other.[22]

_____

[22] This distinction would not be apparent if the user adopted the 'dual band' handsets now coming on the market.

This additional spectrum was provided in most cases for free or for a nominal fee. Only in Germany was this additional allocation of 1800 MHz frequencies subject to an auctioning process. In October 1999, the German regulator organised a simultaneous ascending-bid auction for ten paired frequency blocks. The bidding was limited to the existing four mobile firms which, however, were in an asymmetric position. There were the two GSM 900 firms (T-Mobile and Mannesmann) with a large customer base and capacity constraints on radio spectrum and the two DCS 1800 firms (E-Plus and Viag Intercom) which were still constructing the network and were not constrained by capacity. It was thus clear from the outset that the two GSM 900 firms had a much higher reservation price than the two DCS 1800 firms. The GSM 900 firms had two options: either to bid high immediately to displace the DCS 1800 firms, or to start low and try to share the spectrum. In fact, the auction lasted for only three rounds and the two winners were the GSM 900 firms. This suggests that the first option was chosen. Table 6.6 shows the auction bid for each of the ten frequency blocks. Each frequency block corresponds to 1 MHz except for the last, which corresponds to 1.4 MHz. Mannesmann adopted a jump-start strategy, bidding high for all frequencies and signalling for the blocks 5–10 that this would be the preferred block. In round 2, T-Mobil clearly signalled by closing up on blocks 1–5 that it agreed with this sharing. Since the bids were high the two DCS 1800 firms were not interested in bidding further and thus the auction finished in round 3. The total spectrum auction yielded about DM400 million (€200 million). This may be considered high when compared with past GSM auctions, but relatively low when compared with the UMTS auctions.[23]

6.5 3G auctions in Europe

Simultaneous ascending-bid auctions, as used in Germany, became a popular method of assigning radio frequencies to mobile telecommunications firms in connection with the assignment of frequencies for 3G mobile telecommunications services in 2000–1. The introduction of 3G services was considered a strategic aim of industrial policy at the European level. The development of UMTS networks was strongly endorsed by EU

---

[23] Grimm, Riedel and Wolfstetter (2002) consider this price low, and illustrate the conditions under which a low-price outcome can be the result of a simultaneous ascending-bid multi-unit auction. The crucial role is assigned to the activity rule defining that in absence of a new bid the bidder has to leave the auction. This activity rule tends to exaggerate the strategic demand reduction effect.

Table 6.6 *Highest bids, GSM spectrum auction, Germany (DM million), 2000*

Round	Frequency blocks									
	1	2	3	4	5	6	7	8	9	10
1	36.36	36.36	36.36	36.36	36.36	40.00	40.00	40.00	40.00	56.00
2	*40.10*	*40.10*	*40.10*	*40.10*	*40.10*	40.00	40.00	40.00	40.00	56.00
3	*40.10*	*40.10*	*40.10*	*40.10*	*40.10*	40.00	40.00	40.00	40.00	56.00

Note: The figures in the table indicate the highest bids in each round. Bids in normal font are from Mannesmann, T-Mobil bids are in *italic*.
Source: German regulator (RegTP), as reported in Grimm, Riedel and Wolfstetter (2002).

policies;[24] at a technical infrastructure level, the EU was promoting UMTS for 3G mobile telecommunications services. UMTS could not only provide new services, it could also increase the level of competition in the sector as new companies entered the market. Though the Commission coordinated much of the sector regulation, the decision on the method of assigning licences was retained by the member states. The Commission thus limited itself to indicating that the method of assignment should be non-discriminatory and transparent. Eight out of the fifteen member states selected auctions as the method for frequency assignment. The European Commission, however, criticised the non-coordinated licence allocation mechanisms adopted across the EU.[25] The deployment of UMTS networks was expected to broaden the access opportunities to broadband telecommunications networks and to foster international competitiveness within mobile telecommunications where Europe is at the leading edge in terms of both economic and technological developments. A reasonably fast development of 'European' UMTS was expected to open larger markets in regions outside Europe.

In technical terms, the auction concerned the assignment of a total spectrum range of 2×60 MHz (one range for upstream and one for downstream), as defined by the regulators. To make the service viable, a firm had to have a licence of at least 2×10 MHz. There were thus

[24] See Decision No. 128/1999/EC on the 'Coordinated introduction of UMTS in the Community', as well as European Commission (2001).
[25] The Commission has also raised concerns that excessive licence fees could trigger off collusive market behaviour. Cooperative undertakings such as network sharing arrangements will therefore be closely monitored for their impact on competition.

essentially four options concerning the number and size of licences for 'paired' spectrum:[26]

1. Four equal licences of 2 × 15 MHz each
2. Five licences: two licences of 2 × 15 MHz each and three licences of 2 × 10 MHz each
3. Six equal licences of 2 × 10 MHz each
4. Twelve frequency blocks of 2 × 5 MHz each, with each firm requested to win either two or three blocks. This would lead to four, five, or six licences.

As one of the declared aims of the assignment of 3G licences was to increase competition, most countries made sure that there were at least more licences than incumbents. In principle, all countries declared their aim as being to assign all licences simultaneously. However, in cases of an insufficient number of bidders, the regulator decided to reduce the number of licences to be auctioned in the simultaneous auction and to assign the residual spectrum at a later stage. As can be seen in table 6.7 the first auctions were able to induce a larger number of bidders than there were licences for auction. However, as the number of bidders declined, auctions become shorter and prices lower. In Belgium, the number of bidders was lower than the number of licences, and unsurprisingly the auction lasted only one round and the winners paid the reserve price only. To avoid any risk of collusive behaviour, Denmark decided to change auction design and opted for a sealed-bid auction.

6.5.1 The UK

The UK was the first European country to use the auction method for assigning UMTS licences.[27] At the time, there were already four-facility based cellular mobile telecommunications firms in the market.[28] The aim of the regulator was to increase the competition in the market further by increasing the number of firms by one. This was hardly achievable with only four UMTS licences, as an incumbent firm was perceived as having an advantage in bidding for additional spectrum compared to a new entrant. It was therefore likely that all four licences would attract the highest bids from the incumbents. Five licences were needed, both to avoid the auction being a contest among the existing firms only and at the same time to create

---

[26] 'Paired' spectrum is designed for carrying signals to and from telephone handsets. This contrasts with 'unpaired' spectrum where signals travel only one way. Some licences include also unpaired spectrum, but unpaired spectrum is of much less value than paired spectrum.

[27] Countries such as Finland and Spain had assigned their licences earlier, but through administrative procedures.

[28] Namely Vodafone, BT Cellnet (later mmO$_2$), Orange and One2One (later T-mobile).

Table 6.7 *UMTS auctions, Europe, 2000–2001*

Country	Date	Number of bidders[a]	Number of licences	Number of bidding rounds	Number of bidding days
UK	April 2000	4 I; 9 NE	5	150	52
Netherlands	July 2000	5 I; 1 NE	5	305	14
Germany	August 2000	4 I; 3 NE	6	173	19
Italy	October 2000	4 I; 2 NE	5	11	2
Austria	November 2000	4 I; 2 NE	6	14	2
Switzerland	December 2000	3 I; 1 NE	4	1	1
Belgium	March 2001	3 I;	4	1	1
Denmark	September 2001	3 I; 1 NE	4	1[b]	1

Notes: [a] I = incumbent; NE = new entrant.
[b] Sealed-bid auction.
Source: Author and NAO (2001).

a market structure that would induce sufficient competition. The licences were not identical in terms of size, with two 'large' licences and three 'small' licences. Licence A was the largest, followed by licence B, which was almost the same size as licence A, less the unpaired portion of the spectrum. But unpaired spectrum, as we have seen, is valued much less than paired spectrum. To compensate for the disadvantages new entrants faced, licence A was reserved for new entrants. The further details of licence formats are indicated in table 6.8.

The auction type was simultaneous ascending bidding, working in the following way. In the first round, each bidder made a bid for a licence of choice. According to the activity rule, to remain in the auction each bidder had to either hold the current top bid on a particular licence or to raise the bid on any licence by at least the minimum increment. The auction ended when only five bidders were left. They were allocated the licence on which they were the top bidder, paying the current bid.

The auction started on 6 March 2000 and lasted for seven weeks and 150 rounds, and closed on 27 April 2000. There were thirteen bidders (the four incumbent firms plus nine potential new entrants). This high number of bidders, which was never achieved again in subsequent auctions in Europe, suggested a high degree of competition. Indeed, the receipts from the auction were surprisingly high at £22.5 billion, which was far above the expectations of the UK government.

Table 6.8 *UMTS licences, UK, 2000*

Licence	Paired spectrum (MHz)	Unpaired spectrum (MHz)	Total spectrum (MHz)	Minimum opening bid (£ million)	Winner	Winning bid (£ million)	Price per MHz (paired) (£ million)
A	2 × 15	5	35	35	TIW	4,385	292
B	2 × 15	0	30	30	Vodafone	5,964	398
C	2 × 10	5	25	25	BT	4,030	403
D	2 × 10	5	25	25	One2One	4,004	400
E	2 × 10	5	25	25	Orange	4,095	410

Source: UK Radiocommunications Agency.

Table 6.9 indicates the pattern of bidding for the licences. A widely shared view was that the majority of bidders pursued a strategy of bidding for the licence that represented the best value.[29] The bids thus jumped from one licence to another. Two bidders, however, Cellnet and Orange, deviated from this strategy. Vodafone consistently bid for licence B only. This bidding strategy was interpreted as a determination to win that particular licence. Vodafone also resorted to 'jump bids' – i.e. bids above the minimum increment – to underline this determination. Vodafone's final bid was also a jump bid. Orange followed a similar strategy: initially, it bid for licence B only and once this became too expensive it bid for licence E only. Two bidders determined all prices. NTL was the highest-bidding new entrant that did not receive licence A. This firm determined the prices for licences C, D and E. BT was the highest incumbent bidder not receiving the largest licence B.

The UK Treasury expected substantial receipts from the auction, but the final outcome was far above expectations. Several factors may explain the high prices achieved in the UK auction – the design of the auction, the large number of participants, the fact that this was the first auction in Europe out of a sequence of a relatively large number of countries. This is consistent with the theory that claims that when bidding in a sequence of auctions for complementary items, the early items sell for a higher price. A winner has a competitive advantage in winning subsequent auctions, because it provides a credible signalling that this bidder values them more.[30] Last, but not least, the auction also took place during a stock

---

[29] See, for instance, Cramton (2001), NAO (2001) and Plott and Salmon (2001). For a critical view, see Börgers and Dustmann (2001, 2002).
[30] It remains to be seen, however, to what extent mobile licences in different countries are in practice complementary items.

Table 6.9 *Bidding patterns for 3G licences, UK, 2000*

Bidder	Final bid in round	Number of times bid for licence					Licence won
		A	B	C	D	E	
TIW	131	**12**	5	12	12	10	A
Vodafone	143	–	**34**	–	–	–	B
BT	149	–	23	**8**	13	27	C
One2One	146	–	–	12	**16**	13	D
Orange	148	–	16	–	–	**8**	E
NTL Mobile	145	23	–	12	13	4	
Telefónica	129	5	1	12	20	6	
Worldcom	117	4	–	14	15	7	
One.Tel	97	2	–	9	10	5	
Spectrumco	95	20	1	3	4	1	
Epsilon	94	–	–	13	9	6	
Crescent	90	–	–	8	13	6	
3G UK	89	3	–	8	10	10	
Total bids		69	80	111	135	103	

Note: The figures in the table indicate the number of bids made by the bidders for any licence; **bold** figures indicate the winning bidder for the licence.
Source: UK Radiocommunications Agency.

market bubble that unprecedentedly lowered the financial constraint for firms in the telecommunications sector.

The fact that data for each round of bidding were public induced detailed scientific investigation of bidding behaviour. This real-life experiment may give useful hints about the presumed superiority of auctions as an efficient assignment method for public goods. Studies by Börgers and Dustmann (2001, 2002) came to a more critical valuation of the aptness of auctions as an efficient allocator of radio frequencies. In their first study (2001), they showed that actual bidding behaviour turned out to be inconsistent with the hypotheses implied by the theories of private values and straightforward bidding. According to the private value theory, bidders enter the auction with fixed valuations for each licence, and do not revise their valuation during the auction. Straightforward bidding would suggest that firms bid in each round for the licence for which the difference between the value of the licence and the minimum admissible bid is largest, provided the difference is positive. Once the minimum admissible bids are above the value of each licence, the firm withdraws. Bidders ran against the predictions of these relatively simple rules. The most important inconsistencies arose with the differences in revealed values between large and small licences. For instance, BT switched in round 17 from a large to a small licence, even if prices were to rise much higher later on.

In their second study (2002), Börgers and Dustmann make their benchmark model more flexible by allowing for elements of common value. The auction is thus a learning process by which firms can update their valuation of the licence by observing the action of rival bidders. Moreover, allocative externalities can emerge, depending on which firm is about to receive a licence. However, even this approach cannot rationally explain BT's bidding behaviour concerning the B licence. It is likewise difficult to reconcile NTL's withdrawal from bidding for licence A and the bidding for a small licence, although the absolute level of the price difference between the two licences had shrunk to a relatively low level. The authors conclude that in presence of the poor understanding of the firm's bidding behaviour it is very difficult to trust in auctions efficiently to allocate radio frequencies.[31]

6.5.2 The Netherlands

The Netherlands was the second country to auction UMTS licences. The auction design had several aspects in common with the UK auction, in particular simultaneous ascending bidding for five licences (two 'large' licences for $2 \times 15\,\text{MHz}$ and three 'small' licences for $2 \times 10\,\text{MHz}$). There was, however, one important difference: whereas the UK market had only four incumbent mobile firms, the Netherlands already had five. This difference is crucial if incumbency matters. If incumbents have an advantage, and there is ample evidence that this is the case in the mobile telecommunications market,[32] then Vickrey's revenue equivalence theorem no longer holds and efficiency is not ensured by all auction methods. The chosen set up left few incentives for new entrants to enter the contest, given that there was the same number of incumbents as the number of licences. The contest was rather among the incumbents as to who would receive the large licences. The details of the licences are indicated in table 6.10.

The only non-incumbent bidder was the firm Versatel. However, it publicly declared that it was not interested in having a UMTS licence, but it rather wanted to raise the prices for the other bidders and to extract

---

[31] Professor Börgers bases his conclusions also on his practical experience as adviser to the UK's Radio Communications Agency on the design of UMTS auctions.

[32] For instance, incumbents already have a customer base and it is less demanding to move existing customers to new 3G services than to acquire new customers. Moreover, incumbents have lower fixed and variable costs as they can use a substantial part of the existing 2G network for the building of the 3G mobile telecommunications network, such as base station towers, transmission links and other overheads.

Table 6.10 *3G licences, the Netherlands, 2000*

Licence	Paired spectrum (MHz)	Unpaired spectrum (MHz)	Total spectrum (MHz)	Winner	Winning bid (Guilders million)	Price per MHz (paired) (Guilders million)
A	2×15	5	35	Libertel	1573	105
B	2×15	5	35	KPN	1567	104
C	2×10	5	25	Dutchtone	960	96
D	2×10	5	25	Telfort	948	95
E	2×10	5	25	Ben	870	87

Source: Dutch Telecommunications and Post Department.

concessions for other telecommunications services such as interconnection.[33] The incumbents were thus faced with a free rider problem: all of them would benefit if Versatel dropped out, but at least one party needed to strike an agreement.

The auction had some peculiarities that apparently were not fully understood by the main actors at the time. The government was not entitled to charge a minimum level of licence fees. Moreover, bidders could use 'pass cards' at the beginning of the auction and the prices of licences receiving no bids would be stepwise reduced to zero. One strategy was to use these cards at the very beginning of the game. All firms followed this strategy by bidding the same low price during the initial rounds, with the exception of Libertel. Libertel immediately set a relatively high price, signalling that it was interested in receiving a large licence. Table 6.11 shows the bids at the end of each day of bidding. The strong competition for large licences increased prices relatively fast on them; this can be seen very well in figure 6.1, which plots the price/Hertz of each licence. Large licences are considerably more expensive on a per-Hertz basis than smaller licences[34] (see table 6.10). Prices for small licences were increasing from the initial bid from day 6 of the auction. Towards the end of the auction, prices were increasing fast. The reasons for this became known later. On day 12 of the auction Versatel outbid Telfort on licence D. During the weekend Telfort's lawyers sent a confidential letter to Versatel stating that it was accusing them of bidding only to raise prices and that managers should be made liable for all damages resulting from this. Versatel apparently

---

[33] See Van Damme (2001) for a description of the 3G auction in the Netherlands.
[34] Note that this was the opposite to the situation in the UK.

Table 6.11 *Bidding for 3G licences, the Netherlands, 2000 (€ million)*

Licence date	A Highest bidder	Bid	B Highest bidder	Bid	C Highest bidder	Bid	D Highest bidder	Bid	E Highest bidder	Bid	Total revenues
6.7.2000	Telfort	0.1	Libertel	45.4	Dutchtone	0.1	Ben	0.1	Versatel	0.2	45.9
7.7.2000	Telfort	5.5	Libertel	45.4	Dutchtone	0.1	Ben	0.1	Versatel	0.2	51.3
10.7.2000	Telfort	55.0	KPN	60.4	Dutchtone	0.1	Ben	0.1	Versatel	0.2	115.8
11.7.2000	Libertel	208.7	KPN	208.5	Dutchtone	0.1	Ben	0.1	Versatel	0.2	417.6
12.7.2000	Libertel	252.5	KPN	252.3	Dutchtone	0.9	Versatel	0.8	Ben	0.8	507.3
13.7.2000	Libertel	252.5	KPN	252.3	Telfort	2.0	Versatel	2.0	Ben	2.0	510.8
14.7.2000	Libertel	252.5	KPN	252.3	Dutchtone	5.8	Telfort	5.0	Ben	4.6	520.3
17.7.2000	Versatel	277.8	KPN	252.3	Dutchtone	18.2	Telfort	15.8	Libertel	18.2	582.3
18.7.2000	Telfort	305.6	KPN	305.3	Ben	59.0	Versatel	48.3	Libertel	59.0	777.1
19.7.2000	KPN	406.7	Libertel	406.9	Dutchtone	124.8	Ben	135.1	Versatel	125.8	1199.3
20.7.2000	Libertel	492.1	Telfort	539.1	Dutchtone	222.4	Ben	242.8	Versatel	202.0	1698.3
21.7.2000	Libertel	595.4	KPN	593.0	Ben	358.0	Versatel	390.9	Dutchtone	359.1	2296.4
24.7.2000	Libertel	713.8	KPN	711.1	Telfort	435.6	Telfort	430.0	Ben	395.0	2685.5

Source: Dutch Telecommunications and Post Department.

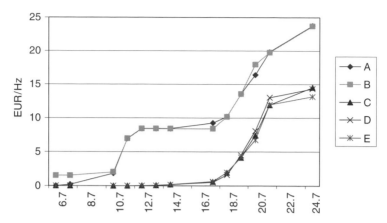

Figure 6.1 *Evolution of bids for UMTS licences, the Netherlands, 2000*
Source: Dutch Telecommunications and Post Department.

interpreted this message as a threat and told Telfort and the auctioneer that it would no longer bid. This message was not transmitted to the other parties and the auction continued the following Monday until it was Versatel's turn to move and it became apparent that the auction was closed. The other bidders complained about Telfort having insider information, and that auction revenues had been higher than necessary.

Nevertheless, total revenues were far less than the government had expected.[35] As a result of the low total revenues and the suspicion of collusion the Dutch parliament started an official investigation into the auction process. Telfort revealed during this investigation that it had accepted an request from Versatel to meet on the day the auction started. Once this became known to the Dutch competition authority, they raided the headquarters of the two companies, but without finding any incriminating evidence.

Several game theorists agree on the fact that the auction's main flaw was its design. Klemperer (2002a) claims that five licences with five new incumbents would have encouraged strong foreign firms to join local incumbents in bidding JVs. This problem was aggravated by the fact that the government could not impose minimum prices, which could otherwise have been taken from the UK experience.[36]

---

[35] Van Damme (2001) reports that the Minister of Finance was expecting revenues in the order of €10 billion.
[36] However, too high minimum prices can create embarrassment also, as the French case showed (see below).

6.5.3 Germany

The German UMTS auction was also an open, ascending, uniform-price auction, but with the important difference that the number of licences was not fixed at the outset.[37] Bidders had to bid for frequency blocks. A total of twelve paired $2 \times 5\,MHz$ blocks were available, so each bidder had to acquire two or three blocks, and depending on the outcome of this the number of licences could vary from four to six. There were paired spectrum blocks on auction; in addition, there were five blocks of 1 MHz each of unpaired spectrum to be sold in a subsequent auction to the winning bidders of the UMTS auction. This second auction could also be used to auction any paired spectrum left over from the first auction.

As an activity rule, bidding rights had to be exercised or would be lost forever: the number of bidding rights in round $n + 1$ are equal to the bids placed in round n. Hence, once a bidder had reduced its bidding rights from three to two blocks, it was impossible to revert to bidding for three blocks. In the auction, only the high bids were made public after every round. Bidders could thus not observe their rivals' bids and did not know precisely their number of bidding rights.

The auction began on 31 July 2000 and lasted 173 rounds of bidding, closing on 17 August 2000. The minimum starting bid was DM100 000 with a minimum 10 per cent increase with each bid (reduced to 5 percent at a later stage). Seven bidders participated: T-Mobil (a subsidiary of Deutsche Telekom), Mannesmann–Vodafone, E-Plus and Viag Interkom, which were all incumbent mobile telecommunications firms; Mobilcom and Debitel were already in the mobile market as virtual mobile network operators (firms that acted as resellers of airtime of incumbent operators, with their own customer base and billing system); finally there was the firm 3G (who assumed the trade name Quam) as truly new entrant.

It is possible to analyse the bidding behaviour of individual firms, focusing in particular on when a firm reduces demand from three to two blocks.[38] Initially all bidders bid for three blocks. After round 115 Debitel reduced demand to two blocks and in round 126 it exited the auction (see table 6.12). From then onward, the auction could have been terminated if the six remaining firms had reduced their demand to two blocks each. Total auction revenues would in that case have been DM61.6 billion. But the auction went on until all remaining firms dropped their demand for the third block. The first to reduce the demand from three to two blocks was Viag Interkom (round 134), followed by 3G (round 138), E-Plus (round 140), Mobilcom (round 146) and

---

[37] For a detailed description, see Grimm, Riedel and Wolfstetter (2001).
[38] Grimm, Riedel and Wolfstetter (2001) describe the method.

Table 6.12 *Critical rounds, German 3G auction, 2000 (DM million)*

Frequency block	Round 126	Bidder	Round 173	Bidder
1	5117.2	E-Plus	8310.4	Viag
2	5129.7	E-Plus	8170.0	Mobilcom
3	4989.0	E-Plus	8330.0	Mannesmann
4	5400.0	Mobilcom	8304.6	3G
5	5203.0	3G	8200.0	Mobilcom
6	5200.0	Mobilcom	8206.6	Viag
7	5368.0	T-Mobil	8304.3	T-Mobil
8	5357.0	3G	8274.3	E-Plus
9	4872.0	Mannesmann	8277.9	Mannesmann
10	4992.1	Viag	8143.9	3G
11	4947.2	T-Mobil	8143.8	T-Mobil
12	4987.3	Mannesmann	8141.4	Mannesmann
Total bids	61,562.5		98,807.2	

Note: In round 126, after the dropout of the seventh bidder, the auction could have been concluded if all bidders had agreed to reduce their demand to two frequency blocks, which eventually occurred only in round 173.
Source: German regulator (Reg TP), as reported in Grimm, Riedel and Wolfstetter (2001).

T-Mobil (round 167). Mannesmann, the last bidder for three blocks, reverted to two blocks only in round 172. The final allocation happened in round 173 when the total auction revenue of DM98.8 billion was reached. The auction thus ended in round 173 with six 'small' licences assigned to all four incumbent firms and two new entrants. Speculation has occurred on why firms insisted on crowding out at least one more rival. The only firm having a vested interest in driving up licence fees could have been T-Mobil, which could have acted in the interest of the German government, its main shareholder. However, this hypothesis is not entirely convincing as T-Mobil dropped its third block bidding right in round 167 and it was Mannesmann–Vodafone which continued to use its third block bidding right until round 173.

It is very puzzling that the bidders apparently paid too much for an outcome that could have been achieved at a much lower level of licence fees.[39] Ewerhard and Moldovanu (2002) rationalise this as an equilibrium outcome when there is a positive probability that incumbents are unable to push potential entrants out of the market. When this pre-emption is unsuccessful, an allocation arises that could have been reached at a lower

[39] The subsequent auction for five unpaired spectrum blocks yielded much lower revenues, a total of DM561 million. Only Viag Interkom did not receive such a licence.

price level if the winning bidders had reduced their demand earlier. *Ex post* regret situations can also have negative welfare implications because of the so-called 'exposure problem'.[40] Or, seen another way, licence fees could have been higher with a less flexible design of market structure, where the number of licences was fixed at a lower level.

6.5.4 Italy

The Italian auction was for five UMTS licences with paired frequency blocks of 2×10 MHz. For new entrants, two additional paired blocks of 2×5 MHz were available. So a new entrant could actually have access to 2×15 MHz spectrum, whereas incumbents could get at most 2×10 MHz. The Italian regulator was keen on having competition among bidders. The auction therefore was for five licences only if there were at least six bidders. To ensure competition among bidders, there was the provision that if there were fewer than six bidders the actual number of licences to be auctioned would be reduced that so that the number of licences would be one fewer than the number of bidders. The rules of the game established that in each round bidders could increase their previous bid. However only the lowest bidder had to increase its bid and the increase should be of at least 5 per cent during the first ten rounds and 2 per cent thereafter. The lowest bidder could pass up to three rounds before increasing the bid. Moreover, each bidder could ask for one day of cessation. The auction would stop once a bidder did not improve on its bid if outbid and when all the passing options of that bidder had already been spent.

Six bidders were qualified to enter the auction after two candidates were discarded because of poor financial credibility (to raise the money necessary to pay for the licence and to build the UMTS network). The qualified bidders were the four incumbent 2G firms Telecom Italia Mobile (TIM), Omnitel (a subsidiary of Vodafone), Wind and Blu, as well as two new entrants, Ipse and H3G. During the pre-qualification stage there were already doubts about the real commitment of some of the shareholders of Blu to finance high licence fees, in particular its minority shareholder British Telecom. Blu was a late entrant in the 2G market and had struggled to acquire customers in a market with an already high penetration rate for mobile telecommunications. These doubts led after the auction to legal battles concerning the right of Blu to participate in the first place. The auction started on 19 October 2000, with a reserve price of Lire 4000 billion

---

[40] This may occur in a multi-object auction with complementarities. Package bids may lead to inefficient non-participation (see Cramton, 1997). As a matter of fact, out of the twelve initially pre-qualified interested firms, only seven participated in the auction.

Table 6.13 *Bids, Italian 3G auction, 2000 (Lire billion)*

Round	Date	Bidders					
		Omnitel	TIM	Wind	Blu	Andala (H3G)	Ipse
1	19.10.2000	4000	4000	4000	4000	4000	4000
2	19.10.2000	**4230**	*4000*	4220	4210	4200	*4000*
3	19.10.2000	**4250**	4220	4220	4210	*4200*	4220
4	19.10.2000	4250	4220	4220	*4210*	**4430**	4220
5	20.10.2000	4250	*4220*	*4220*	**4440**	4430	*4220*
6	20.10.2000	4250	*4220*	*4220*	4440	4430	**4450**
7	20.10.2000	4480	*4440*	4470	**4490**	4460	4450
8	20.10.2000	4480	**4680**	4470	4490	*4460*	**4680**
9	20.10.2000	*4480*	4680	**4700**	4490	**4700**	4680
10	21.10.2000	**4740**	4680	4700	*4490*	4700	4730
11	23.10.2000	**4740**	4680	4700	*4490*	4700	4730

Note: The highest bids are in *bold*, the lowest bids in *italic*.
Source: AGCOM data.

(€2.1 billion). Table 6.13 shows the bids in each round. From this it emerges that TIM, the largest 2G firm, had a very cautious approach, having bid the highest bid only once. Omnitel, on the other hand, had bid the highest bid four times. On 21 October, after round 10, Blu asked for a day of cessation, in line with the auction rules. However, when the auction resumed on 23 October, Blu failed to increase its offer. The auction thus finished at round 11, with each winning bidder paying on average Lire 4700 billion (€2.4 billion). Although the auction yielded slightly more than the government's reserve price, it was far below market expectations. This led to heated political discussions and Blu was accused of participating in a plot by having entered the auction without a willingness or capability seriously to bid. However, an investigation by the Competition Authority (AGCOM, 2001) led to the conclusion that there was not sufficient proof of collusive behaviour to invalidate the assignment of the licences. Italy was thus the first case where an incumbent 2G firm was not able to successfully bid for a 3G licence during an auction.

6.5.5 Austria

The Austrian auction design was similar to that in Germany. There were twelve paired $2 \times 10\,\text{MHz}$ frequency blocks on auction, with each participant bidding for two or three blocks. Six participants entered the auction: Mobilcom (the mobile subsidiary of fixed line incumbent Telekom Austria), Maxmobil, Connect Austria and tele.ring, which all had already a 2G licence, as well as Hutchinson 3G and 3G Mobile, which were new entrants. The minimum bid for a frequency block was €50 million. The auction could, in principle, have been concluded immediately if the bidders had agreed to bid for two blocks each, avoiding the aggressive bidding for three blocks by some bidders as observed in Germany. Mobilkom apparently made statements ahead of the auction that it would not bid aggressively if others would do the same (Klemperer, 2002a). This message seemed to have been observed as the auction ended after 16 rounds with each bidder receiving two blocks and paying on average €118 million per block. The number of bidders not exceeding the maximum number of potential licences and the hint about a 'soft' bidding strategy by the dominant firms are elements that support the relatively early conclusion of the auction.

6.5.6 Switzerland

In Switzerland, four UMTS licences were put on auction. There were three 2G firms already in the market and the auction should have helped to accommodate at least one new entrant.[41] Initially ten applicants qualified for the auction. However during the run-up to the auction there was a continuing dropout of qualified bidders and eventually only four bidders showed up on 13 November 2000, when the auction was supposed to start. The four bidders were the three incumbent 2G firms Swisscom, Orange and diAX, plus the new entrant Team 3G (a subsidiary of Telefónica). At this stage, the Swiss regulator called off the auction and wanted to postpone it to an unspecified future date. However, following firm lobbying and public pressure, the regulator eventually agreed to undertake the auction on 6 December 2000, with four bidders for four licences. Unsurprisingly, the auction was concluded immediately, with each of the four bidders paying the reservation price of SF50 million, apart from Orange which paid an extra SF5 million to ensure a particular block of frequencies to match those in a neighbouring country.

This auction was the most successful in Europe until that date and a public debate on the design of auctions followed. It is clear that an exact matching of the number of licences with the number of bidders does not induce competition. To increase bidder participation in such circumstances, Klemperer (2000), for instance, proposed switching from an open,

---

[41] For a detailed description of the auction, see Wolfstetter (2001).

ascending-price auction to a one-time sealed bid. This would partially mitigate the disadvantage new entrants had in bidding, since the incumbents might not bid their maximum valuation and so the new entrants' bid could be higher than the incumbent's bid. This turned out to be the case in Denmark, as will be seen shortly. However, the sealed-bid auction had at least two drawbacks. First, it might deviate from efficiency, as a weak bidder would have a positive probability of receiving a licence. Second, it would expose bidders to the 'winner's curse' to a larger extent.

6.5.7 Belgium

In Belgium, four UMTS licences were scheduled for auction, which meant that there would be one more licence than the number of incumbent 2G firms. However, as only the three incumbents expressed interest in obtaining a licence, the auction was called off and a licence was given to each of the three incumbents at the reservation price of €150 million on 2 March 2001.

6.5.8 Denmark

Denmark was in a difficult position to organise an auction in face of the mounting failure to attract a sufficient number of participants. The plan was to assign four UMTS licences, which was exactly equal to the number of incumbent 2G firms, namely TDC Mobile (a subsidiary of former fixed line monopolist TeleDanmark), Sonofon, Telia Mobile and Orange. To make the participation attractive for new entrants, the regulator adopted the sealed-bid auction, with all winning bids paying the fourth-largest bid. This incentive mechanism seemed to work, as in fact the firm HI3G participated too. The sealed-bid auction was made on 19 September 2001 and the incumbent 2G firm Sonofon was outbid. Each winner paid €125 million.

6.6 Discussion of the experience of European 3G auctions

Although governments do not claim that raising revenue is the main reason for selecting auction as the assignment mechanism, the size of the licence fee has generally been seen as a benchmark of the success of the auction. Figure 6.2 indicates licence fees, in *per capita* terms, in EU countries. The licence fees are listed in chronological order according to the time of assignment, and the figure shows the huge disparity of the licence fees determined during the European auctions. The range was from €630 per head in the UK to €44 in Belgium. This would suggest that the

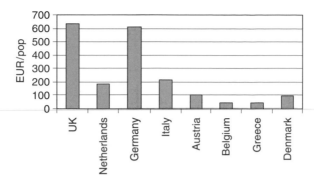

Figure 6.2 *Licence fees, European 3G auctions, 2000*
Note: The countries are listed in chronological order of auctions.
Source: European Commission data.

revenue raising capability of auctioning is mixed, and also that the variance in valuations is not plausible.

The wide variation in licence fees may be attributed to several elements. A major determinant of the licence fee is the design of the licence. As Klemperer (2000) pointed out, the design of licences matters crucially. General theoretical statements such as the 'equivalence theorem' do not apply in practice. The design of the auction thus needs to take into account real-world phenomena and practices. First of all, the design should encourage competition, so the number of the bidders should strictly be larger than the number of licences available. Moreover, if participants have different private values (e.g. incumbent vs. new entrant), disadvantaged bidders must nevertheless be induced to bid by providing them with a positive probability of submitting a winning bid. Last but not least, the scope for collusive bidding and signalling should be kept as restricted as possible.

Another fundamental issue is the fact that licences were awarded sequentially across Europe. Such an approach gave early auctions a head start as firms that were aiming to obtain multiple licences put a decreasing valuation on licences over time. Firms learning about auctioning behaviour tend to bid less aggressively in subsequent auctions. This hypothesis, however, has difficulty explaining the low result in the Netherlands and the increases observed in Denmark and Ireland towards the end of the sequence.

Financial constraints could also emerge among bidders. At the beginning, this constraint might not be so great and become significant only as more and more licences were acquired by bidders with multinational strategies. One has also to take into account the fact that the auctions for European UMTS licences started when the 'new economy' stock market bubble of the late 1990s started to deflate. Access to financial markets for bidders became increasingly difficult, with the cost of funding rising sharply because of a rapid decline in stock market valuation of telecommunications firms.

This sequential licensing in Europe was mainly due to the absence of a regulator who could have coordinated the process. A single European-wide auction process for national licences, like the allocation of PCS licences in the USA, could have mitigated this distortion in the licensing process across countries.

As will be discussed in more detail in chapter 7, it became clear in the aftermath of the auctions that there had been extensive overbidding and miscalculation of the licences' value. Several firms have in fact handed back their UMTS licences or postponed the setting up of their UMTS infrastructure, completely writing off the licence fee in their accounts.

6.7 Administrative procedures for European 3G spectrum licences

Several countries in Europe maintained the tradition of relying on administrative procedures. The main reason for this method was purportedly related to the public interest in having an early and widespread access to the new type of services. Administrative procedures are thus believed to give regulators more leeway in influencing the development of the mobile telecommunications market in the context of a technological environment that is evolving very fast. As will be seen with the examples of Finland and Sweden, administrative procedures may become quite sophisticated.

6.7.1 Finland

Finland led the way in assigning 3G licences in March 1999, a year ahead of Spain which was the second country in Europe to assign 3G licences. The reason for this early assignment was indicated in the following quote from the Finnish government:

The Ministry of Transport and Communications has accelerated the process of granting the licences, because Finland also wishes to be a pioneer in the implementation of third-generation mobile technology. By granting the licences as early as possible, Finland will signal the other countries of the fact that licences can be applied for and handled promptly, which again will enhance the implementation of mobile telecommunications in other countries. After the early granting of licences, the licensees have more time to concentrate on the development of networks when they are certain that they can commence commercial activities.[42]

---

[42] http://www.mintc.fi/www/sivut/english/tele/telecommunications/index.html (the website of the Finnish Ministry of Transport and Communications).

Industrial policy considerations were clearly the predominant consideration in assigning the licences. The Finnish Ministry of Transport and Communications awarded four national licences to the four 2G firms Radiolinja, Sonera, Telia Mobile and Suomen Kolmegee (the latter comprising forty-one regional fixed line telecommunications companies of the Finnet group). There were twelve candidates entering the 'beauty contest' for the licences. The Ministry provided a very detailed public statement as to why the winners were chosen and why the losers were rejected.

6.7.2 Spain

There were six candidates participating in the 'beauty contest' for four licences taking place in March 2000. The priorities were declared to be those of rapid network deployment and development of the market. The licence fee of €25 million per licence was therefore set at a low level. The winners were three incumbent 2G firms Telefónica, Airtel and Amena, as well as the new entrant Xfera.

Following the high licence fees achieved in auctions held in the UK and Germany, pressure mounted on the Spanish government to revise the terms of the 3G licences granted. The government yielded to public pressure and the licence fee was retroactively increased to €130 million and a supplementary annual tax equivalent to 0.015 per cent of net 3G revenues was imposed. However, as the firms faced increasing difficulties in obtaining access to financial markets in the light of the stock market collapse after 2000, the Spanish government had again to backtrack and ease the licensing conditions in terms of network build-out and licensing period. This illustrates the problem of time consistency of regulatory obligations and the reputation of policy makers. This precedent could lead to opportunistic behaviour of candidates in future licensing processes.

6.7.3 Sweden

The Swedish government had delegated the Swedish regulator PTS to assign 3G licences. The regulator opted for a 'beauty contest' for four licences, one more than there were 2G incumbents. The argument was as follows:

In contrast to many other European countries, Sweden and the other Nordic countries do not hold auctions to award mobile telephony licences. Swedish law stipulates that licences must be allocated based on specific criteria. This is to the advantage of operators and consumers alike, because the operators do not have to pay the state expensive fees for licences. (National Post and Telecom Agency, quoted in Hultén, Andersson and Valiente, 2001)

The applicants were supposed to submit their applications by September 2000, paying an application fee of SK100 000 (€17 000). These fees actually did not cover the cost of the 'beauty contest', which were assessed at more than two person-years of work. The selection process was a two-step procedure. In the first step, an evaluation was made whether the applicants had fulfilled the preconditions for the establishment of a network in accordance with the plans presented in the application, by focusing on financial capacity as well as technical and commercial feasibility. Those candidates passing this first hurdle were then evaluated on the detailed merits of their proposals in the second step.

There were ten participants in the 'beauty contest', among which were the three incumbent 2G firms. Table 6.14 indicates the different participants and the key features of their proposals. All candidates presented plans that in their own evaluation gave a good population coverage (100 per cent for most candidates), but the differences of territorial coverage were much larger. Territorial coverage between the smallest (Broadwave) and the largest (Mobility4Sweden) was 1:12. In terms of investment, cost proposals also varied widely between the lowest budget of SK6.8 billion proposed by Telia (an incumbent firm) and the largest one of SK36.9 billion proposed by HI3G (a new entrant). These huge differences also reflected the large technological uncertainty involved in 3G technology.

Five candidates dropped out at stage one:[43] four competitors failed on technical feasibility and one failed on financial capacity. Two applicants (Broadwave and Tenora) failed because of patent mistakes in their proposals. Broadwave would clearly have been unable to cover the claimed area with the base stations indicated in the application. Tenora could not pass either the financial capacity test or the technical feasibility tests. Mobility4Sweden made a mistake in financial capacity considerations, in particular omitting to indicate the funding of start-up losses. Reach Out Mobile AB was excluded because of failure to pass the technical feasibility test. However, the most surprising result of the first step was that Sweden's former telecommunication monopolist was excluded. Telia failed because of failure to pass the technical feasibility test. One important reason for this was that Telia committed itself to build only 4100 base stations, whereas according to the regulator it would need to construct three times more base stations in the countryside than in its application to fulfil the coverage requirements.[44]

Thus five consortia passed the criteria in phase 1: Europolitan, HI3G, Orange, Tele 2 and Telenordia. In phase 2 the applicants were evaluated on

---

[43] For a more detailed description see Hultén, Andersson and Valiente (2001).

[44] Telia, however, entered the 3G market through a JV with the licence winner Tele 2.

Table 6.14 *Key figures for applications for 3G licences, Sweden*

Applicant[c]	Status[b]	Base stations[a]	Costs (Kronor billion)	Area covered (km²)	Kronor million per km²	Kronor million per base station	km² covered per base station	Population covered (per cent)
HI3G	NE	20184	36.9	224 724	164	1 828	11	100
Europolitan	I	20000	27.5	165 259	166	1 375	8	100
Tele 2	I	10186	17.7	112 666	157	1 738	11	100
Mobility4Sweden	NE	8760	15.3	395 520	39	1 747	45	100
Orange Sweden	NE	8635	19.7	364 528	54	2 281	42	100
Tenora	NE	7550	11.2	290 038	39	1 483	38	100
Telenordia Mobil	NE	7200	14.0	181 346	77	1 944	25	98
Reach Out Mobile	NE	5238	15.8	259 944	61	3 016	50	100
Broadwave	NE	4700	14.7	32 750	449	3 128	7	81
Telia	I	4100	6.8	304 661	22	1 659	74	100

Notes: [a] The base stations refer to the number of base stations foreseen in phase 1 (end 2003).
[b] I = Incumbent, NE = New entrant.
[c] The winners are in **bold**.
Source: Author's elaboration on PTS data, quoted in Hultén, Andersson and Valiente (2001).

area and population coverage as well as roll-out speed with a weighted marking system. Telenordia had proposed population coverage of 98 per cent at the end of 2003. The other four competitors proposed that nearly 100 per cent of the population should have access to 3G services at the end of 2003. Europolitan, HI3G, Orange and Tele 2 were therefore assigned the four licences in December 2000. The commitments made in the applications entered the licence conditions. PTS (the Swedish telecommunications regulator) was in charge of monitoring the build-out of the 3G networks; the most important date was December 2003, since the four winners promised to complete their networks at that date. The provision was that if a firm had not performed in accordance with the commitments it might be subject to a fine – or, in the extreme case, the licence could be revoked altogether. As the licencing concerned a new technology with much of the basic equipment still under development at time of licensing, such a threat was not credible. Delays in implementing networks are relatively easy to justify and by December 2003 all the licence winners had failed to meet their commitments. Only the new entrant HI3G had started restricted services.

After the declaration of the winning candidates, legal proceedings started, as the three losers challenged the assignment decision in the courts. However, as the regulator had issued the licences in one decision, it was difficult for the challengers to ask for revocation of the assignment as a whole, as they would have to prove that a mistake was made in all four winning bids. In any case, the county administrative court decided in June 2001 that the assignment decision was correct and rejected most of the complaints; although it made some criticisms, it did not affect the appropriateness of the ultimate decision.

6.7.4 France

France organised a 'beauty contest' for four 3G licences. This meant there was one more licence than there were 2G operators. The French regulator also suggested a fee of €0.6 billion per licence in March 2000. This was overruled by the French government, which also in the light of the high licence fees paid in the UK, increased the fee to €5 billion. The actual assignment procedure was planned to be held six months later. However, as time passed and licence fees obtained in European auctions declined it became increasingly clear that such a licence fee would be difficult to achieve. Eventually, only two bids (those by the two largest incumbent 2G firms, France Télécom and SFR) were filed. The third 2G firm (Bouygues Télécom) did not bid at all. France Télécom and SFR were assigned the licences in July 2001, with the provision that the remaining licences would be assigned at a future date. At this stage however pressure mounted on the government to reduce the licence fee, and in October 2001 it was restored to €0.6 billion plus a turnover tax on 3G services of 1 per cent.

In December 2001 a second 'beauty contest' was organised for the two remaining licences, on the same terms. Only Bouygues Télécom participated and was eventually awarded the licence. From a regulatory point of view the French case can be considered either as bad luck for the latecomer or poor timing. By yielding to pressure to reduce the bid, the French government has lost reputation and reneged on the time consistency of the announcement of conditions.[45]

6.8 Conclusion

This chapter has looked at issues arising in the management of radio spectrum, the key input for mobile telecommunications services. As the spectrum licence is the barrier of entry for a facility-based firm in this oligopoly market, assignment procedures are important and are often intertwined with political considerations. A major distinction has been made between administrative procedures such as 'beauty contests' and market-based procedures such as auctions. Over time, a shift towards market-based systems could be observed. The justifications turn on arguments such as reducing the administrative burden or complexity, as well as improving the ability to identify the firms that will make the most efficient use of the spectrum. The auction design is of crucial importance for this latter condition to be satisfied. In the context of the 3G auctions in Europe, the assumption of rational behaviour of bidders has been questioned, in particular in those countries where the auction receipts have been largest. However, administrative procedures have also shown their weakness in terms of governments' ability to enforce the commitments bidders made at the time of the licence assignment. Chapter 7 will explore in more detail the extent to which apparently inconsistent behaviour by firms reflects their strategic behaviour.

---

[45] On these issues, see Morand, Mougeot and Naegelen (2001).

7 The evolution of market structure in mobile telecommunications markets

7.1 Introduction

As seen in chapter 6, market entry in mobile telecommunications is based on a licensing process for scarce spectrum resources. Whereas initially the assignment of licences was based on administrative procedures, there is a trend towards market-based mechanisms such as auctions. All European countries assigned radio spectrum for the provision of 3G mobile telecommunications services around the years 2000–1. The licence assignment procedures varied across countries, but the majority made recourse to auctions. Auctions were an innovative method and produced several surprising results. Unexpectedly high licence fees were observed in some countries, whereas in others they were far below expectations. Some of these outcomes, especially the disappointing ones, were explained to some degree by differences and inconsistencies in auction design. But several points still beg an explanation, especially the high licence fees. However, the administrative procedures also for licence assignments, such as 'beauty contests,' produced disappointing results on several occasions. The economists' typical argument in favour of an auction revolves around the assertion that it is a market-based mechanism, and hence best to select the agent with the highest willingness to pay (WTP). That would implicitly ensure that the most efficient use of a scarce public resource was made. However, the track record emerging from 3G licensing in Europe sheds some doubt on whether performance in attracting efficient firms has improved through auctions. For this purpose a close look is now taken at the actions of governments and firms in the aftermath of the 3G auctions in Europe. The first 3G auctions in Europe, in particular in the UK and Germany, led to licence fees that were far above expectation. This induced several observers and governments to benchmark the success of a licence assignment by the size of the fee.[1] The fee appeared to be determined by the financial resources available rather than by the intrinsic value of the object at stake. This means also that little concern was expressed in Europe on the compatibility of the fees with the envisaged market structures. This chapter abstracts from public finance considerations and looks at the implications of the size of licence fees for the evolution of market structure.

Mobile telecommunications technologies come in generations, and with each new generation governments have the opportunities to design market structure anew by assigning new radio spectrum licences. The observed trend suggests that governments increased competition in the market by assigning an increasing number of licences as new generations of technologies were introduced. For instance, chapter 4 proved that increasing the number of firms was typically conducive to a faster diffusion of mobile telecommunications services. As Dana and Spier (1994) have shown in a model of auctions and endogenous market structure, the government's incentives to increase or decrease the number of firms depend on the amount of information available to them. Incomplete information induces a bias toward less competition. Given the fast technological change and the high uncertainties of market prospects for the mobile telecommunications, market valuations by firms and governments may diverge strongly.[2] Governments wanted to continue the trend of increasing the number of firms in the mobile telecommunications industry with 3G, but at the same time they allowed for a mechanism to escalate licence fees. The key question, then, is whether the chosen market structure, as given by the number of licences to be assigned, is consistent with the licence fee raised. Suppose that industry profits fall with the number of firms in the market; there appears to be a trade-off between the number of firms in the market and the licence fees that can be paid out from expected oligopoly rents. 'Overbidding' will occur if a firm engages to pay a higher licence fee than the expected oligopoly profit. It is argued that overbidding may lead to a much more concentrated industry or encourage collusive market behaviour. If auctions encourage overbidding, problems of time consistency of regulatory policy may emerge, especially with respect to the regulatory commitment to enforce competition and rapid diffusion. If high licence fees lead to higher levels of concentration and prices, a slowdown in

---

[1] See for instance Binmore and Klemperer (2002). The total spectrum fees for 3G frequencies as collected by governments over 2000–1 exceeded €100 billion (European Commission, 2002a). In comparison, the spectrum auctions organised in the USA over 1994–2001 yielded some \$40 billion (Cramton, 2002).

[2] As will be seen later, for GSM services both firms and governments initially made cautious market growth assumptions, but actual growth vastly exceeded initial expectations. For UMTS on the other hand, the high licence fees could discount the expectation of a very rapid market growth, with firms being far more optimistic than governments.

diffusion of new services may be the consequence, with adverse welfare effects. This line of reasoning is contrary to the traditional argument that licence fees are sunk costs and thus should not affect post-entry behaviour, but it follows recent experimental research on the sunk-cost fallacy (Offerman and Potters, 2000). To illustrate the arguments, a theoretical framework is now presented, focusing on the interplay between market structure and endogenously determined fixed costs.[3] The features of the model are then contrasted with the empirical evidence from the European mobile telecommunications industry.

The chapter, based on Gruber (2004), is organised as follows. Section 7.2 presents the theoretical model to analyse the market structure as a function of licence fees. Section 7.3 presents background information on the mobile telecommunications industry in Europe. Section 7.4 describes the design of market structure for the 3G markets across European countries and the results of the licensing procedures. Section 7.5 makes a critical assessment of the outcomes and comments on the developments subsequent to the auction. Section 7.6 draws some conclusions.

7.2 The theoretical framework

7.2.1 The model

Consider a homogeneous goods industry with Cournot competition[4] and with the following inverse demand function, $p(Q) = s/Q$. s is a parameter for market size and Q are total quantities sold at price p. Assume constant marginal costs, c. It can be shown that in a Cournot equilibrium with n (with $n > 1$) identical firms (where quantity supplied by each firm is $q = Q/n$) the equilibrium price is $p = nc/(n-1)$. As typical for a Cournot model, price is above marginal cost and declining with the number of firms. The fixed entry cost, F, sets an upper bound on entry. At a Cournot equilibrium the profits for each firm are:

$$\Pi(n, s, F) = (p - c)q - F = s/n^2 - F \qquad (7.1)$$

The Cournot equilibrium number of firms n^* is determined by the following zero entry condition:

$$\Pi(n^*, s, F) > 0 > \Pi(n^* + 1, s, F) \qquad (7.2)$$

---

[3] This approach has similarities with the endogenous sunk-cost literature. For instance, in Sutton (1991, 1998) 'sunk costs' refer, respectively, to advertising expenditure and R&D costs.
[4] As already seen, the Cournot competition assumption is supported by empirical evidence for the mobile telecommunications industry.

Neglecting the integer problem, from (7.1) one can derive the following expression:

$$n^* = \sqrt{\frac{s}{F}} \tag{7.3}$$

One can thus derive relationships between equilibrium number of firms, market size and entry costs: $dn^*/ds > 0$ and $dn^*/dF < 0$. The equilibrium number of firms thus increases with market size and decreases with fixed costs.

7.2.2 Regulatory failures

Let us now relate this model to an industry such as mobile telecommunications. Using the above comparative static features, observations can be made on the impact of changes in exogenous variables (such as technology or policy changes) on the equilibrium number of firms. To start with, suppose there were no spectrum constraint. As seen above, with free entry, the Cournot outcome would be n^* firms and zero profits.[5] If, however, entry is regulated and the number of firms set at $\tilde{n}$, then three cases are possible:

1. $\tilde{n} > n^*$: This case implies excessive entry and negative profits.
2. $\tilde{n} = n^*$: This case corresponds to the free-entry outcome with zero profits.
3. $\tilde{n} < n^*$: In this case, regulated entry is less than the free-entry outcome with positive profits. These 'oligopoly rents' decrease as the number of firms increases.

To make the issue of licence fee explicit, it may be useful to redefine the fixed cost, F, as follows: $F = I + L$. This means that the fixed cost is split into network investment[6] costs, I and licence fee, L. Equation (7.3) can therefore be rewritten as:

$$\hat{n}^* = \sqrt{\frac{s}{I + L}} \tag{7.4}$$

i.e. $\hat{n}^*$ defines the equilibrium number of firms when a licence fee, $\hat{L}^*$ is involved. Likewise, n^* defines the equilibrium number of firms when a licence fee is zero. Abstracting again from the integer problem, we have $\hat{n}^* < n^*$ for $\hat{L}^* > 0$ – i.e. with licence fees we should have a smaller equilibrium number of firms than with zero licence fees, as licence fees

---

[5] Let us abstract here from the integer problem.

[6] Network investment costs covers a wide range of costs, including the costs related to uncertainties in technological development such as delays in the availability of suitable equipment. A further analysis of this is not scope of this study, but the qualitative effects on changes in these costs on equilibrium outcomes are comparable to changes in the licence fee. Greater uncertainty in technological development thus compound the effects deriving from an increase in the licence fee.

are equivalent to increasing fixed entry costs. $\tilde{n}$ is fixed by the regulator and L could be decided by either the regulator or by the firm. It is important that the licence fee, L, is consistent with the total number of firms that are supposed to coexist in the market. According to the 'policy' variables, L and $\tilde{n}$, a series of relationships between variables n^*, $\hat{n}^*$ and $\tilde{n}$ are possible. The most interesting cases of regulatory failure are shown in boxes 7.1–7.3.

Box 7.1. Excessive entry: $\tilde{n} > n^* > \hat{n}^*$

Here, the regulator provides more licences than the equilibrium number of firms at a zero licence fee. This is a case of excessive entry and thus not a stable market structure even with a zero licence fee.

Box 7.2. Excessive licence fee: $n^* > \tilde{n} > \hat{n}^*$

In this case, the licence fee has been set at such a high level that an otherwise viable market structure (i.e. with zero licence fee) becomes unstable. In this case, the competitive equilibrium would lead to $n = \hat{n}^* < \tilde{n}$.

Box 7.3. Excessive profits: $n^* > \hat{n}^* > \tilde{n}$

In this case, the licence fee is low enough that all firms can coexist with non-negative profits. A licence fee has extracted only some of the oligopoly profits.

7.2.3 Endogenous licence fees

Assume now that firms decide about the licence fee to pay. Firms therefore play a two-stage game. In the first stage, they decide the licence fee and in the second stage they compete *à la* Cournot. Licence fees can thus be considered as an endogenous sunk cost. The Nash equilibrium outcome is as above and (7.4) may be rewritten as:

$$\hat{L}^* = s/n^2 - I \qquad (7.5)$$

where $\hat{L}^*$ is level of licence fee that drives industry profits to zero. The iso-profit relationship between the number of firms (market structure) and the

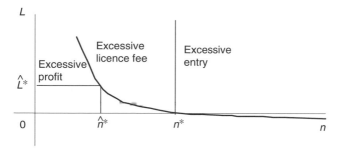

Figure 7.1 *The iso-profit relationship (with zero profit)*
Note: The curve is the graph of the zero profit combinations of the number of firms (market structure) and licence fee. Right from the curve, industry profits are negative, left from the curve, industry profits are positive.

licence fee is illustrated in figure 7.1. The curve is the graph of the zero profit combinations.[7] Right of the curve industry profits are negative (due to excessive entry or an excessive licence fee, i.e. box 7.1 and 7.2); left of the curve industry profits are positive (excessive profits, i.e. box 7.3). Thus for any $\tilde{n} = \hat{n}^*$ chosen by the regulator, we need $L = \hat{L}^*$ to have an equilibrium. Otherwise licence fees have post-entry effects: if $L > \hat{L}^* \geq 0$ we have excessive entry. With $L < \hat{L}^*$ we may have a stable market structure, but excessive profits may create problems of resource allocation. Both cases may thus impair the efficient working of the market.

There are many ways to determine the size of the fee. If there is competition among firms for the licence, the size of the fee offered by an individual firm becomes a determinant for spectrum allocation. Competition for spectrum licences increases the licence fee and thus may endogenously affect market structure. In principle, a higher licence fee tends to reduce the equilibrium number of firms in the industry. The government typically determines how many licences will be granted and thus exogenously sets the number of firms in the industry. If firms set the licence fee, the endogenously determined licence fee might become incompatible with the exogenously set market structure. If, for instance, firms are paying too high licence fees, exit of some firms may be necessary to re-establish non-negative profits. As will be seen later, auctions are most likely among the several allocation mechanisms to produce 'excessive entry' and thus the highest probability that some firms may actually exit after having been allocated a licence.

---

[7] This curve would, of course, shift with changes in the key variables. For instance, an increase in I (that could be due to higher infrastructure costs following increased technological uncertainty) would shift the curve to the left. Likewise a reduction in infrastructure cost (that could be due to infrastructure sharing among firms) would shift the curve to the right.

By fixing the number of licences at the outset, the government sets the market structure exogenously. The traditional argument in favour of an auction is that it allocates the spectrum to the most efficient firm which will value it most. The validity of this proposition is, however, based on two premises. First, the government does not license too many firm (i.e. avoids the situation in box 7.1). Second, firms do not collude once they have entered the market. This second point is developed in more detail below.

7.2.4 Post-entry effects

Suppose that the number of licences is set at $\tilde{n}$, and that the zero profit condition is fulfilled with a licence fee of at most $\tilde{L}$, given the technology. If firms bid $L > \tilde{L}$, then there are negative profits in the industry, unless some exit occurs. If industry profits rise as the number of firms declines, it is easy to show that a monopoly will pay the highest licence fee, as a monopoly has the highest rents to dissipate.[8] A tension therefore emerges between the objective of extracting high licence fees from spectrum assignment and having as many firms as possible in the industry.

Any specific licensing policy for the underlying services needs to be justified. The policy maker typically wants to assign the licences to the firms that are best at diffusing the associated services. There are two major decisions to be taken: the number of licences to be allocated and the determination of the licence fee. If prices are driving market growth, then the effect of the number of licences essentially depends on the type of competition. If Bertrand competition were prevailing, then two firms would be enough to establish competitive prices. If instead Cournot competition were prevailing, then the price is a decreasing function of the number of firms. If price is a determinant for market growth, this will increase with the number of firms. From this it follows that in order to have the largest number of firms in the market, the licence fee should be zero.

The second question relates to the post-entry effects of licence fees. Economic theory would suggest that up-front sunk cost should not interfere with post-entry competition, as pricing decisions are based on marginal costs. But what if excessive licence fees have been paid? Suppose, for example, that in a duopoly framework the duopoly profit is less than the licence fee paid. In that case, there are two options for the firm: *exit* or *collusion*. With the exit of one firm, the remaining firm could reap monopoly profits and thus break even. If, on the other hand, the government can credibly commit itself to a

---

[8] As argued by Grimm, Riedel and Wolfstetter (2003), if a bidder could get all the spectrum, the winner of the auction would monopolise the market.

duopoly structure, then firms need to collude to reap monopoly profits to repay the licence fee. High licence fees can therefore lead to higher prices than without a licence fee. As such, licence fees can be seen as an inducement to collusive behaviour. Moreover, market growth will be lower.

The question then arises whether competitive auctions for licences provide incentives to establish escalating licence fees. Put in another way, can auctions for licences induce credible signalling for collusion at the post-entry stage? Imagine the case of an auction for two licences. If post-entry collusion is ruled out, auctioning with firms with identical cost structures will lead to licence fees that drive profits to zero. Licence fee L will be equal to the duopoly firms' profit $\Pi(2) = L$. In other words, licence fees perfectly extract all oligopoly rents. But we know also from the previous discussion that the Cournot duopoly firm's profit is less than half of the monopoly profit $\Pi(1)$: $\Pi(2) < \Pi(1)/2$. From this, one can derive an excessive licence fee that will be profitable with collusion as long as it is in the range $\Pi(2) < L < \Pi(1)/2$. In other words, spectrum allocation through auctions can lead to extraction of monopoly profits with collusion, and not necessarily to the allocation of scarce resources to the socially best use.

7.2.5 International aspects

This model also permits analysis of the consequences of licence fees for international shifting of oligopoly profits. Consider two countries with different policies for assigning licences. A country that establishes a small licence fee benefits from the fact that firms have less of an incentive to collude. With Cournot competition, the largest possible number of licences need to be granted to have low prices and the highest levels of diffusion. Oligopoly rents will not go to the government, but will rather be shared among consumers and producers. Prices will be low and market penetration high. In the country that chooses high licence fees, possibly determined through competitive auctions, there may be an inherent incentive for firms to collude. In that case monopoly prices will be charged. Most of the rents will go to the state via the licence fee. However, there will be high prices and therefore lower market penetration.

Differences in the licensing regimes across countries can have implications for firm performance. To illustrate this point, consider the following simple framework. Suppose two identical countries denominated 1 and 2. Each country has one firm, called firm 1 and firm 2, respectively, but each firm can operate reciprocally in both countries. Finally, suppose that each firm has a cost advantage when operating on the domestic market compared to when operating in the other country (e.g. lower marginal costs due to information advantages). With Cournot competition, this leads to higher market shares for the domestic firm on the domestic market.

Assume now that country 2 establishes a licence fee and thus extracts some oligopoly profit. This puts firm 2 at a disadvantage because it has a larger market share in a low-profit market (country 2) and a small market share in a high-profit market (country 1). For firm 1, the reverse holds: it has low market share in a low-profit market and high market share in a high-profit market. Hence firm 1 has higher total profits than firm 2. Not asking for licence fees can also be seen as a subsidy to firm 1, especially from country 2's perspective. Legitimate questions now arise as to whether the two firms are now forced to compete on unequal terms, whether the absence (or, in any case, inequality) of licence fees are distorting subsidies and whether coordination of the regulatory frameworks within countries participating in a common market is desirable.

Finally, there are also issues concerning lump-sum transfers of rents. Firms active in country 2 pay a higher licence fee than in country 1, for two reasons: first, because there is an auction which drives up the licence fee; second, because there is no licence fee in country 1, firms have more funds available to spend on a licence in country 2. Firms active in country 1 can thus employ some of the forthcoming rents from country 1 in country 2 to bid for the licence.

7.3 The profitability of the mobile telecommunications sector in Europe

The cellular mobile telecommunications industry became the first major laboratory of competitive supply of telecommunications services in a sector where the natural monopoly paradigm was prevailing. In many countries, however, this opportunity for competition was appreciated only after some delay. Initially, most countries viewed mobile telecommunications as just an additional new business of the state-owned telecommunications monopoly, the development of the cellular network was a means of honing the innovative capabilities of national equipment suppliers. In the early days of mobile telecommunications, licences were typically granted on a first-come-first-served basis, if not automatically, to the incumbent fixed line telecommunications operator. A few countries granted a second licence, which was assigned through an administrative tender procedure (or 'beauty contest'). With the introduction of digital technology based on the GSM standard, the European Commission started actively to promote a coordinated approach with more competition. Member countries were instructed to grant at least three licences for digital services. There were generally three or four firms in each European national 2G mobile telecommunications market,[9] with a relatively stable market structure.

It is fair to say that during the 1G and also at least during a large part of the 2G technology, the mobile telecommunications market was in an 'excessive profits' situation. These technologies were far more successful in the market than originally expected and produced huge oligopoly rents. Table 7.1 lists the profitability of selected European mobile telecommunications firms in 1997, a period of high growth in the mobile telecommunications market. It shows that for some firms, such as TIM, profitability in ROCE-terms could be even above 100 per cent. Other firms such as Telecel, Vodafone and Mannesmann had a ROCE that was still several multiples of the typical industry average of around 15 per cent. Profitability declined rapidly in the following years, however, mainly as a result of enhanced competition in the market and also because of the expense of acquisitions and licences. For TIM for instance, ROCE declined by 15.8 per cent by 2001. For the other firms in table 7.1 it was difficult to find comparable data because they became involved in mergers and acquisitions (M&As), and hence only consolidated data is available. However, the decline in profitability is illustrated by industry data from the UK regulatory authority. Table 7.2 shows the ROCE for the four firms in the UK mobile telecommunications market as calculated by the regulator Oftel (2002). Apart from the large difference in the level of profitability, all firms with the exception of Orange showed a decline in profitability. Oftel calculated an average cost of capital of 15 per cent for mobile firms, all firms other than Vodafone had profits below this level from 1999 onwards. That means that only Vodafone remained profitable. The lack of profitability of One2One (later called O2) and Orange is to

Table 7.1 *Profitability of selected European mobile telecommunications firms, 1997*

Firm	Country	ROCE[a]
TIM	Italy	137.1
Telecel	Portugal	65.5
Vodafone	UK	54.0
Mannesmann	Germany	41.9
Comvik	Sweden	17.9
Netcom	Norway	14.2

Note: [a]ROCE is the percentage return on capital employed.
Source: Firm accounts, listed in Warburg Dillon Reed (1998).

[9] The exceptions were the Netherlands with five firms and Luxembourg with two firms (see also table 7.3.

Table 7.2 *Profitability (ROCE) of mobile telecommunications firms, UK, 1998–2002*

	1998	1999	2000	2001	2002
Vodafone	92	76	53	50	45
BT Cellnet	20	12	8	c	n.a.
One2One	−18	−5	−23	c	n.a.
Orange	5	5	10	n.a.	n.a.

Notes: ROCE is the percentage return on capital employed
c = Confidential
n.a. Not available.
Source: Oftel data

a large part also explained by their later entry compared to Vodafone and BT Cellnet (later called O_2).

Regulatory activity in the mobile telecommunications industry has been reduced compared to the fixed line telecommunications industry, as the sector is considered as liberalised and competitive. Regulatory action in most cases is thus limited to market segments such as interconnection, where there is scope for abuse of market power. Firms with 'significant market power' (i.e. with market shares typically in excess of 25 per cent) are subject to separate accounting in order to establish cost-based pricing in interconnection.[10] There is the presumption that regulatory action or entry will drive profitability towards the average level of the economy; as already shown for the UK, there is evidence that later entrants in the market have much lower rates of return.[11] There has even been a case of exit in Italy, where in 2002 the smallest and latest (fourth) entrant Blu left the market.[12] In most European countries, licence fees were either zero or relatively modest, especially when compared to what would be paid for 3G licences.[13] The struggling for survival of fourth entrants in some countries

[10] This was the case with regulation in place in the EU until 2003. With the new (2002) regulatory framework, these requirements were further relaxed, with the aim of bringing telecommunications regulation closer to competition law.

[11] For instance, Oftel (2002) showed for the UK industry that the early entrants BT Cellnet and Vodafone had a big competitive advantage because they could operate a less expensive GSM network at the 900 MHz frequency range compared to the later entrants One2One and Orange, which had to operate a more expensive GSM network working at a 1800 Mhz frequency range.

[12] The spectrum was sold on equal terms to the three remaining firms.

[13] Total licence fees collected in the EU for 2G licences amounted to €10 billion, whereas for 3G licences it was, as already mentioned, in excess of €100 billion (European Commission, 2002a).

suggested that a three- or four-firm market structure came close to the zero profit entry condition in the industry. More than four firms were thus likely to constitute 'excessive entry' in the 2G market.

A widely followed business growth strategy for external growth was internationalisation through acquisition of existing firms and bidding for new licences.[14] The best example for this was the astonishing growth of Vodafone, which developed from a UK firm to be the largest internationalised mobile telecommunications firm worldwide. Orange and T-mobile were other examples. Most of the firms adopting this expansion strategy experienced a rapid deterioration of their profitability and credit rating. Investor sentiment about those firms shifted from optimism to deep pessimism, creating substantial financial difficulties, compounding the burden deriving from financing the acquisition of 3G licences in the wake of the bursting of the financial 'bubble' in 2000.

Technical difficulties also hampered the introduction of 3G technology, since it involved the establishment of a new technology on a very large scale. Whereas the 1G and 2G mobile telecommunications systems were mainly designed for voice transmission, the next technological step was the development of systems for data transmission. 3G systems substantially increased data transmission rates and permitted the sending of moving images. These services were provided on one of the five competing internationally defined technology standards. The EU member states committed themselves to introducing 3G under UMTS, a concept developed by ETSI. The European interest was in making UMTS backward-compatible with the existing installed GSM base.[15] The first adoptions of 3G systems occurred in 2002 in Europe. In Japan, it happened somewhat earlier, though with a slower than expected adoption of 3G services by users. In Europe, launch dates, officially set for January 2002, were delayed by more than a year. The first commercial launch was undertaken by subsidiaries of the multinational firm Hutchinson, under the label 3, in Italy and the UK during the second quarter of 2003 (ITC, 2003a). Equipment problems, especially the limited availability of 3G handsets, were one of the main reasons for the delay in introducing 3G services. The incumbent firms in the market were also not particularly keen in promoting 3G services, as they had invested in upgrading their 2G networks to

---

[14] For an extensive account of such diversification strategies, see Whalley and Curwen (2003).
[15] European policy makers were very keen on introducing early 3G systems for reasons of industrial policy (European Commission, 2001). Early adoption of UMTS was seen as crucial for preserving the worldwide lead in mobile telecommunications technologies established with GSM (see European Commission, 1997a, 1997b). EU member states were thus instructed to provide 3G licences so that the first 3G technology-based services could become available by 2002. For the expected evolution of 3G mobile telecommunications, see Gruber and Hoenicke (1999, 2000).

deliver services such as multimedia messaging, that 3G were supposed to provide. Only the new entrant 3 had a genuine interest in advancing diffusion of 3G services.

Other countries, such as the USA delayed the development of 3G systems because of the slow development of the 2G systems that were introduced late and used a range of different, non-compatible technologies (ITC 1999). On the one hand, competition among different 2G systems slowed down diffusion compared to a market with an established technology standard (Gruber and Verboven, 2001). On the other, competition among 2G systems had provided the opportunity of establishing CDMA[16] as one of the mobile telecommunications systems in the US market during the 2G era. Because 3G technologies were based on the working principle of CDMA, firms already having a CDMA system for 2G services did not need to acquire additional radio spectrum, as 3G services could be provided by simply upgrading the current system. A standard might therefore be helpful in quickly diffusing a given technology, but it might delay the emergence of a new, superior technology.[17]

7.4 The design of market structure for 3G markets in Europe

The entry pattern for the 3G market had a completely different design from previous technology generations. With 1G and 2G markets the evolution of the market structure emerged from a sequential licensing of new entrants, typically starting either with a monopoly (for most 1G services) or with a duopoly (for 2G services). For the 3G service industry the design of the market structure entailed the simultaneous entry of a relatively large number of firms (typically four–six). Little attention seemed to be devoted to the zero profit entry condition in the design of market structure. The $n+1$ rule of thumb (with n being the number of incumbent 2G firms) was typically applied to determine the number of 3G licences. This rule of thumb had a two-fold purpose: to create more competition at both the pre-entry as well the post-entry stage. At the pre-entry stage, new entry would be encouraged to enter the competition in the market; at the post-entry stage, new entry should increase competition in the market. In this game, the incumbents were presumed to have a strategic advantage. Without increasing the number of licences, pre-entry competition for licences would have been weak. The additional licences thus gave the new entrants incentives to bid. This would also help to improve the terms on which governments assigned licences, in particular increasing

---

[16] Code division multiple access.

[17] On this trade-off between standard setting and competition among systems, see also Shapiro and Varian (1999).

licence receipts when combined with an auction process. For the post-entry stage it was expected that additional entry would increase competition, leading to lower prices and better service. However, there was little exploration of whether the market would accommodate $n+1$ firms in a competitive setting.

Concerning the allocation method, half of the EU countries opted for a market-based mechanism such as auctions and the other half for a 'beauty contest'. Italy adopted a hybrid approach, using a 'beauty contest' first, followed by an auction (see table 7.3). In general, the multiple round ascending-price auction was chosen, with the exception of Denmark, which opted for a sealed-bid auction.[18] There was substantial variety in the outcomes across countries, only in part explained by the different assignment method used. The most striking differences could be observed within the group of countries that organised auctions. Figure 7.2 shows the evolution over time of the auction receipts across the different countries, demonstrating a pattern of decline. There is a growing literature trying to rationalise these results, with explanations relying on arguments of bad auction design, collusion and political interference (see Klemperer, 2002b, 2002c and Cramton, 2002). There is, however, the indisputable fact that auctions have, on a *per capita* basis, yielded much more than 'beauty contests', as can be seen from table 7.3. This lists the countries in the order of *per capita* licence fees. The UK and Germany have a *per capita* licence fee in excess of €600, far above other countries. But auctions are not a guarantee of high fees. In the cases of Greece and Belgium, the auction was considered disappointing, the licence fee was low because there were fewer bidders than licences. It was even lower than in the 'beauty contest' in Ireland. As might have been expected, 'beauty contests' were much more prone to political interference. In Spain and France, the licence fee was repeatedly modified following the success and failure of auctions in other countries. The French government raised the licence fee proposed by the regulator by a factor of three, only to cut it to one-ninth after poor success in attracting bidders.

In the cases where firms were allowed to shape market structure themselves, the results were particularly surprising. Germany and Austria auctioned frequency blocks instead of single licences, which to a certain degree allowed market structure to be determined endogenously during the auction. By using frequency blocks any market structure up to the maximum of six firms becomes possible, allowing market structure to become an

---

[18] A sealed-bid auction is a more efficient design when the number of licences to auction is equal to the number of incumbents. The Netherlands did not stick to this, in fact using a multiple-round ascending-price auction for five licences with five incumbents. The result was a short auction,with fees falling far below the government's expectations.

Table 7.3 *3G licence assignment, EU*

Country	Incumbent firms	3G licences planned	3G licences granted	3G licences not assigned	Assignment method[b]	Licence fee/population (€)
UK	4	5	5	0	A	634
Germany	4	4–6	6	0	A	615
Italy	4	5	5	0	BC+A	212
Netherlands	5	5	5	0	A	186
Austria	4	4–6	6	0	A	101
Denmark	4	4	4	0	A	96
Ireland	3	4	3	1	BC	92
Greece	3	4	3	1	A	45
Belgium	3	4	3	1	A	44
Portugal	3	4	4	0	BC	40
France[a]	3	4	3	1	BC	21
Spain	3	4	4	0	BC	13
Finland	3	4	4	0	BC	0
Sweden	3	4	4	0	BC	0
Luxembourg	2	4	3	1	BC	0

Notes: [a]Initially only two licences were assigned; a third licence was awarded to the third incumbent during a second tendering in 2002.
[b]A = Auction; BC = 'Beauty contest'.
Source: European Commission (2002a).

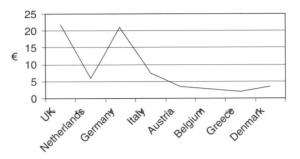

Figure 7.2 *Licence fees, 3G auctions, chronological order (€/head/5MHz)*
Source: European Commission (2002a).

outcome of the licence process itself. In both cases, the least concentrated market structure emerged.

Views on the 3G market structure vary widely. Looking at all countries, the number of 3G firms ranges from three to six firms (see column 4 in table 7.3). In four countries (i.e. the Netherlands, Denmark, Greece and Sweden) the regulator did not contemplate an increase in the number of licences. This means that in those countries an increase in the number of mobile telecommunications firms would occur only if at least one of the incumbents did not receive a 3G licence. It turned out that in all cases except Greece one incumbent did not receive a licence. What is also striking is that in three cases of administrative procedures (France, Ireland and Luxembourg), the number of licences eventually granted was smaller than the planned number.[19] With auctions, the maximum number of licences granted was always achieved, when the number was endogenously determined (Germany, Austria).

A substantial part of the critique of European 3G licensing concerned the sequential assignment process across European countries. Sequential licensing may have strong implications for bidding behaviour in auctions and they are not yet fully understood from a theoretical point of view. However, theory predicts that winning bids in sequential auctions should have a declining price profile because of the reduction of risk and learning effects. This would be in line with the empirical observation as shown in figure 7.2. An example of decreasing prices in multiple objects in sequential auctioning is the 'afternoon effect' (McAfee and Vincent, 1983), which can be explained by the willingness of risk-averse bidders to pay a risk premium at an early stage. In that case, simultaneous auctioning will be more efficient than sequential auctioning. The high licence fees derived from the

---

[19] Notice that in Luxembourg this happened with a zero licence fee.

(early) auctions in the UK and Germany apparently left countries with later assignment dates in an unfavourable position. Several bidders, typically the local subsidiaries of multinational firms, faced increasingly difficult access to finance and thus had to revise their bidding strategies, in particular towards demand reduction. Late-coming countries found it increasingly difficult to attract bidders for 3G licences as the licence process unfolded. It thus appears that the bidding strategy by firms was a function rather of the financial resources available than the intrinsic value of the licence at stake. The drying up of financial resources also had implications for countries that did not organise auctions, but set minimum fees for 'beauty contests'. For instance, the 'beauty contest' in France had to be deferred because only two firms were interested in the contest for the four licences, which however came with a relatively high minimum licence fee. This high fee reduced the demand for licences below the number of licences on offer. To attract further firms in a second round of bidding, the government had to lower the minimum fee. This price reduction was also retroactively granted to the two firms that had already received the licence at the higher fee.

7.5 The aftermath of 3G licensing

The 3G auctions delivered mixed results. Taking licence fees as a parameter of success, some auctions were very successful but others were complete failures (Cramton, 2001; Klemperer, 2002a). Even for apparently well-designed auctions, there is evidence that bidders' actions were not always consistent with rational decision making (Börgers and Dustmann, 2002). Moreover, the view is spreading that 'successful' auctions are delivering licence fees that are far too high for the revenues expected to be generated by 3G services.[20] The renegotiations of licence conditions from assignments achieved by 'beauty contests' suggest weaknesses in this assignment method (as well as of governments).[21] But what is more worrying is the perception by some firms that they may have been subject to the 'winner's curse', i.e. to have paid to much for their licences.[22] Financial concerns arose not only from the burst of the speculative 'bubble' in the financial markets, but also

---

[20] European Commission (2002a) indicates some countries where this could be the case. The study also mentioned that some network build-out proposals presented even in 'beauty contests' with zero licence fees might be too ambitious to be supported by the market.

[21] The European Commission (2002b) has criticised the non-coordinated licence allocation mechanisms adopted across the EU. However, the Commission could not do very much about this as licensing was a prerogative of member states and the only aspects it could enforce were transparency of the process and non-discrimination.

[22] Some firms, such as Telefónica and KPN in fact fully wrote off the in 3G licence fee, setting the accounting value of the licence equal to zero.

the sobering thought that the whole market potential for 3G services might be much lower than expected, whereas both investment and operating costs might be much higher. Gruber and Hoenicke (2000) elaborate on the question of whether the speed of adoption proposed and the size of required investments were supported by a sufficiently high level of demand. In a simulation exercise they showed that revenues from new data services would have to increase by a very large proportion to make 3G services profitable. It is unclear which type of 3G applications requiring large data transmission could generate these revenues. Firms are also exposed to a high risk of introducing the new technology too early. Detailed technical surveys suggest that the investment for a 3G network infrastructure is much higher in comparison with a 2G infrastructure, and operating costs are likely to be higher as well (European Commission, 2002a). If this is so, then a less concentrated market structure than with 2G services may not be supported in the 3G market and exit will be necessary.[23] Thus the $n+1$ rule, the prerequisite for multiple-round ascending price auctions to work, may have turned out to be inappropriate.

This suggests that several European countries are now in a situation of 'excessive entry'. Competitive market equilibrium should therefore be restored through exit. In principle, there are alternatives to exit. These could be, for instance, collusion, or cost reduction through measures such as 'softening' the terms of licensing conditions (e.g. 'softer' conditions for licence fee payments and network investment) and in sharing costs. However these *ex post* changes in licence terms could be accommodated to only a very limited extent as otherwise the licences granted by the regulators could be legally challenged by firms that did not win licences or did not apply in the first place. The reputation and credibility of governments was thus at stake. Ultimately, high licence fees could undermine the time consistency of regulatory policies.

A number of events after the 3G licensing[24] showed how difficult it may become for governments to enforce the terms on which the licences were assigned. There has been a general trend towards delaying the build-out of networks and the supply of 3G services. The main reasons are purportedly technical difficulties and non-availability of equipment, in particular handsets. Moreover, increasing scepticism has been raised about the market potential of 3G services. Several firms that received a licence decided to postpone the building of the network infrastructure, thereby flouting the

---

[23] The study for the European Commission (2002a) also reports findings from a simulation exercise that market structures in Germany, the Netherlands, Sweden and the UK are unlikely to be able to support all firms.

[24] Such events are regularly reported in the trade press and on specialised websites, such as www.totaltele.com.

regulatory commitments. This means that they risk losing their licence when the regulator checks on the implementation of the licence conditions.[25] Other firms even decided to hand back their licence to the regulator, forgoing the licence fee paid (as in Norway).[26] With the justification of reducing costs, several firms also started to build networks on a shared basis with their competitors. The German Chancellor has called on operators to cooperate in building-out the networks.[27] National regulators are observing such schemes with close interest and also with apprehension. The European Commission (2001) has expressed concern about network sharing, as this may be a potential means of collusion. However, most national regulators encouraged network sharing, in particular when it was deemed to accelerate the pace of introduction of 3G services. In several countries the terms of the licence obligations were eased[28] and regulators signalled flexibility in interpreting licence obligations, in particular on the timing of the start of the service and the extent of 3G network coverage.

These adverse market developments precipitated situations that had initially not been considered. What happens if a firm does not want to exercise the rights entailed by its licence. Can such a licence be sold in its entirety to another new entrant or can only the spectrum be sold? In the first case market structure would be maintained, but in the second case there would be a higher level of concentration with an increase in the inequality of spectrum distribution. In some countries firms have already made official statements that they will forgo the licence. In Germany, for instance, the firms Mobilcom and Quam decided to abandon the building of a 3G infrastructure. Mobilcom sold the 3G network infrastructure it already had in place to existing 3G firms, but was unable to sell the licence because of lack of interest; no firm was interested in acquiring the licence because of its attached network build-out obligations. At the same time existing 3G licence holders were not entitled to buy the spectrum. In 2003 the firm thus decided to return the licence to the regulator, forgoing the upfront licence fee of €8.4 billion with little hope of receiving any compensation. In Germany two firms had already thus exited the 3G market. In many other countries, such as Italy, Austria, Sweden and Portugal, firms, typically the new entrants, announced that they would not be building 3G network infrastructure. In Spain, the spectrum can be traded,

[25] The 3G subsidiaries of the Spanish firm 'Telefónica' did this in Germany, Austria and Italy, where they were all new entrants.
[26] In September 2003, Norway awarded this licence, after a second 'beauty contest', to Hutchinson, at zero licence fee. For the second licence that became available, no firms expressed interest.
[27] See the German newspaper *Frankfurter Allgemeine Zeitung* (14 March 2002).
[28] For instance, in Italy and Sweden the length of licences was renegotiated, extending the licence period from fifteen to twenty years.

and in principle the whole licence can be sold to another firm. The new entrant firm Xfera decided to sell part of the spectrum it had been assigned to one of the other 3G firms. Adjustments to the market structure as designed by governments are thus widespread.

A further issue, not yet fully resolved, has emerged from the reuse of licences that have been returned. The first case of this kind occurred in Norway, where two firms out of 400 gave back their licence. When the government auctioned the two licences again in September 2003, only one bidder expressed interest. A licence was eventually awarded to the firm 3 at a significantly lower price than at the previous auction. However, the existing licence holders were compensated by a 'softening' in network build-out obligations.

The reported anecdotal evidence provides support for the excessive entry hypothesis in several countries. High licence fees exacerbated this situation, as for instance in Germany and Austria. By auctioning frequency blocks, these countries chose a mechanism for determining market structure endogenously. The outcome was that six firms, the maximum, were awarded a licence. In Germany, the least concentrated market structure was combined with exceptionally high licence fees.

There is a broad consensus among economists about the advantage of auctions as a means for allocating scarce public resources. However, auctions may lead to surprising outcomes because some underlying behavioural assumptions may be violated in practice. In-depth studies on the consistency of the bidding behaviour in Germany and the UK[29] have cast doubt on the assumption of rationality of firm behaviour. The postulate that the auction process is a more efficient allocation mechanism of a scarce resource may thus require qualification.

7.6 Conclusion

This chapter has analysed the relationship between radio spectrum licence fees' allocation and market structure in the mobile telecommunications industry. A theoretical framework illustrated the regulatory trade-off in designing market structure for 'natural oligopolies' and the scope for extracting oligopoly rents through licence fees. The number of firms a competitive market is able to support decreases as the licence fee increases. The market structure as designed by the regulator may be overridden by firm behaviour, in particular when entry costs can be determined endogenously. It has been shown how firms might have an incentive to 'overbid'

---

[29] See Grimm, Rieder and Wolfstetter (2001); Börgers and Dustmann (2002).

for licences because they could induce more relaxed competition in the post-entry stage.

The empirical relevance of this result was illustrated with reference to the assignment of 3G mobile telecommunications licences in the EU and by contrasting the outcomes derived from 'beauty contests' and auctions. Auction receipts varied widely across countries – in some countries, far above expectations. They are also likely to be incompatible with economically viable operations in the industry. The exit of new entrants from the 3G market already observed supports the hypothesis of excessive entry. There are also increasing calls for leniency in antitrust enforcement, 'softening' of regulatory obligations and outright calls for subsidies. All this suggests that in several cases the design of the market structure prior to the assignment of the licences was inadequate. The widely adopted $n+1$ rule for 3G licences, while effective in creating competition for entry into the 3G market, seems not to have led to a viable competitive market structure. This problem of excessive entry is exacerbated when market structure is determined endogenously, the cases of Austria and Germany show that the auctions led to the least concentrated market structures.

Auctions have traditionally been justified as an efficient means for putting an economic value on a scarce resource and for allocating it to the firm that will use is most efficiently. While auctions have introduced a more efficient allocation of the spectrum in the USA, because they relieved the FCC of what was considered too high an administrative burden in assigning the very large number of licences, a more qualified judgement may be necessary in the light of the European experience. Many governments have stated that the policy objective for assigning a licence was to introduce new mobile telecommunications services at low prices. According to the model presented, this would be achieved by allowing the highest number of firms to enter the market. Demanding licence fees, however, would reduce the number of firms the market could support in a competitive equilibrium, thus increasing prices and reducing the speed of diffusion. Gauging the success of a licence assignment procedure by the revenues raised may thus be misleading. It has been shown that auctions may induce excessive entry compared to the fee paid, this could then impair the government's ability to enforce the licence terms. Licence fees could prove counterproductive in promoting quick diffusion of a new technology, either because of a more concentrated industry or because of the incentives for collusion.

The lesson to be drawn for the design of market structure is that the choice of the licence allocation mechanism has crucial importance for *post-entry performance*. The issue can be put as starkly as that the regulator must determine whether there should be competition *for* the market or competition *in* the market. This may also require a rethink on the recourse to 'market-based' allocation mechanisms for public goods. Finally, in the light of the observed exit of firms one may also conclude that spectrum is no longer a scarce economic resource for the provision of mobile telecommunications services. These questions all provide scope for stimulating further research.

Appendix

A1 Radio spectrum as a scarce resource

The main economic value of the radio spectrum lies in its capacity to carry information. The radio spectrum is part of the electromagnetic spectrum, which comprises all electromagnetic waves that are transmitted through space. However, not all the spectrum is usable for the purposes of mobile telecommunications. We shall now describe the technical characteristics of the radio spectrum.

A1.1 Classification of radio spectrum

There are two alternative measurement units for the radio spectrum. Hertz (Hz) indicates the number of wave cycles per second. The alternative is to express radio frequencies in terms of wavelength, which is the ratio between the speed of light through a vacuum (approximately 300 million metres per second) and the radio frequency. The wavelength of a 3 kHz radio wave is thus 100 km, while the wavelength of an 300 GHz radio wave is only 1 mm. Table A1 lists the classification of radio frequencies by names of the waves and their wavelength. In the following, Hertz will be used as a measure.

Radio frequencies are a natural resource. It was only during the course of the first half of the twentieth century, along with development of wireless communications, that radio frequencies became also an economically valuable resource. Wireless communications exploit the properties of the radio spectrum to transmit signals. In fact, before 1930 the radio spectrum was empty of manmade signals. The portion of the spectrum that is technically usable for telecommunications depends on the state of technology. Technological progress also means that the range of usable frequencies is widening. As will be seen later, at the beginning of the industry mobile telecommunications used VHF signals, and gradually migrated to

Table A1. *Classification of radio frequencies*

Wavelength	Frequency	Frequency type
1–10 mm	30–300 GHz	EHF: extra high frequency
10–100 mm	3–30 GHz	SHF: super high frequency
100–1000 mm	0.3–3 GHz	UHF: ultra high frequency
1–10 m	30–300 MHz	VHF: very high frequency
10–100 m	3–30 MHz	HF: high frequency
100–1000 m	0.3–3 MHz	MF: medium frequency
1–10 km	30–300 kHz	LF: low frequency
10–100 km	3–30 kHz	VLF: very low frequency
100–1000 km	0.3–3 Hz	MF: medium frequency

Note: 1 GHz = 10^9 Hz, 1 MHz = 10^6 Hz, 1 kHz = 1000 Hz.

UHF signals. Radio frequencies up to 60 GHz are currently considered as operationally feasible, whereas frequencies up to 300 GHz have been tested in laboratories.

A1.2 Technical limits of radio frequencies

The reason for this limited suitability of radio frequencies is that the spectrum does not have homogeneous properties across the whole range of frequencies. The propagation laws, background noise and reflection or absorption characteristics affect frequencies differently. As a general rule, a freely propagating signal undergoes attenuation as a function of distance and frequency. Attenuation increases with the distance travelled and the frequency of the signal.

The range of a radio communications link is defined as the furthest distance that the receiver can be from the transmitter and still maintain a sufficiently high signal-to-noise ratio (SNR) for reliable signal reception. The SNR received is degraded by a combination of two factors: beam divergence loss and atmospheric attenuation. Beam divergence loss is caused by the geometric spreading of the electromagnetic field as it travels through space.

As the original signal power is spread over a constantly growing area, only a fraction of the transmitted energy reaches a receiving antenna. For an omnidirectional radiating transmitter, which broadcasts its signal as an expanding spherical wave, beam divergence causes the received field strength to decrease by a factor of $1/r^2$, where r is the radius of the circle, or the distance between transmitter and receiver.

Beam divergence loss happens to all frequencies. There are causes for the decline of the SNR which are related to atmospheric attenuation and thus specific to individual frequency bands. They depend in particular on the propagation mechanism, or the means by which the signal travels. Radio waves are propagated by a combination of three mechanisms: atmospheric wave propagation, surface wave propagation and reflected wave propagation.

In atmospheric propagation the electromagnetic wave travels through the air along a single path from transmitter to receiver. The propagation path can follow a straight line, or it can curve around edges of objects, such as hills and buildings, by ray diffraction. Diffraction permits cellular telephones to work even when there is no line-of-sight transmission path between the radiotelephone and the base station. Atmospheric attenuation is not significant for radio frequencies below 10 GHz. Above 10 GHz under clear air conditions, attenuation is caused mainly by atmospheric absorption losses; these become large when the transmitted frequency is of the same order as the resonant frequencies of gaseous constituents of the atmosphere, such as oxygen, water vapour and carbon dioxide. Additional losses due to scattering occur when airborne particles, such as water droplets or dust, present cross-sectional diameters that are of the same order as the signal wavelengths.

Surface propagation applies to low radio frequencies, when terrestrial antennae radiate electromagnetic waves that travel along the surface of the Earth as if in a waveguide. The attenuation of surface waves increases with distance, ground resistance and transmitted frequency. Attenuation is lower over seawater, which has high conductivity, than over dry land, which has low conductivity. At frequencies below 3 MHz, surface waves can propagate over very large distances. Ranges of 100 km at 3 MHz to 10,000 km at 1 kHz are not uncommon.

Sometimes part of the transmitted wave travels to the receiver by reflection off a smooth boundary. When the reflecting boundary is a perfect conductor, total reflection without loss can occur. However, when the reflecting boundary is dielectric, or of a non-conducting material, part of the wave may be reflected while part may be refracted through the medium – leading to a phenomenon known as 'refractive loss'. When the conductivity of the dielectric is less than that of the atmosphere, total reflection can occur if the angle of incidence (the angle relative to the normal, or a line perpendicular to the surface of the reflecting boundary) is less than a certain critical angle. Common forms of reflected wave propagation are ground reflection, where the wave is reflected off land or water, and ionospheric reflection, where the wave is reflected off an upper layer of the Earth's ionosphere.

The effects of these different propagation mechanisms change with the frequencies. In the frequency band between 0 and 30 MHz, diffraction or ionospheric reflection allows the signal to be transmitted over very long distances without significant diffraction. The reflection intensity is variable, too, and may depend on particular atmospheric conditions or solar activity. These frequencies are used for intercontinental telegraphy, communications with submarines and radio broadcasting.

In the frequency band above 30 MHz and up to 3 GHz, effects deriving from reflection become negligible and surface attenuation, due to background noise, becomes predominant. The VHF and UHF bands, thanks to their good reflection on walls and good penetration inside buildings, and in spite of their high sensitivity to natural barriers (e.g. woods, valleys), are very suitable for telecommunications applications.

A1.3 Evolution of technological constraints

There is a trend toward the use of higher frequencies in mobile telecommunications. The advantage is that the size of the blocks indicated in table A1 is on a logarithmic scale – i.e. the higher-frequency block contains almost ten times more bandwidth (i.e. frequency range) than the previous one. Bandwidth is equivalent to transmission capacity (as will be seen later), and hence transmission capacity is increasing over time.

There are two countervailing factors that limit the range of the frequencies that can be used for mobile telecommunications. First, signal attenuation increases with frequency and distance. Second, beam loss decreases and sensitivity to interference from manmade noise such as electrical engines, car ignition and domestic appliances decreases with frequency. From optimising these countervailing effects it turns out that with currently available technology the suitable range for mobile telecommunications is within 0.4 and 2.5 GHz.

A1.4 Institutional organisation of radio spectrum

The spectrum suitable for use in wireless communications' applications is divided into frequency bands. There are ranges with upper and lower limits used for particular applications such as terrestrial radio and television as well as mobile telecommunications. In this respect, the spectrum is also an international resource, very much like international flight routes or geostationary satellites. Every nation has sovereignty over the spectrum used within its territory, with the proviso that it should not create interference with the spectrum outside its territory. However, as certain frequencies have propagation properties that give them reach beyond national territories, frequency use is harmonised on an international scale through the International Telecommunications Union (ITU). The ITU for this purpose organises the World Radio Conferences (WRC) which are in charge of allocating frequencies and reorganising the use of frequencies in the light of technological changes and new forms of use.

The 1987 WRC had the special aim of assigning frequencies for mobile telecommunications services. The document that incorporates the international agreements on the use of the spectrum is the Radiocommunications Regulations. This has the properties of an international treaty. Chapter 8 of this document contains the worldwide attribution of frequencies to the type of wireless services.

Frequency bands are subdivided into channels, the frequency units at which communications stations transmit or receive. The term 'channel' thus refers to the means for unilateral transmitting of signals between two points. The size of channels depends on the applications and on the technology standard. A television signal is transmitted on channels with a bandwidth of 6 Mhz, while a mobile telecommunications system such as GSM uses 25 kHz.

A2 The working principles of cellular telecommunications systems

This section describes the basic working principles of cellular mobile systems. These are basic descriptions, and apply to both analogue and digital systems. The differences between analogue and digital technologies are examined in chapter 2.[1]

A2.1 The basic concept

In a cellular system, a large service area is divided into smaller areas or cells. Within each cell, a subsystem similar to a multichannel trunked system is operated with a low power base station in the middle. This layout makes it possible to reuse the frequencies for different mobile users. The base station transmitters in adjacent cells operate on different sets of frequencies to avoid interference. The size of the cell is determined by the transmission power of the base station: the smaller the cell, the more often the frequencies can be re-used. The fundamental advantage of the cellular approach is therefore that more users can be accommodated than with any other mobile communication technology, as long as the users are not concentrated in one cell most of the time. If an existing cell has reached capacity limits, it can be further subdivided into cells according to the same principle, but with lower-power transmitters. The increased scope for frequency thus obtained also permits an increase in traffic handling capacity.

---

[1] For more details see Redl, Weber and Oliphant (1995), Calhoun (1988) and Webb (1998).

There are other advantages to cell-splitting. It allows one to spread the investment cost of countrywide cellular systems along with the growth in traffic. The system starts off with rather large cells and as customers and traffic increase the size of the cells decrease along with the increase in their number. New cells can be created without scrapping the existing investment in the large-radius cell site equipment; the power of those transmitters would simply be scaled down to fit within the new system. Of course, cell-splitting can be applied in a geographically selective manner: small cells for traffic-intensive urban areas and large cells in more suburban and less densely used areas.

Cellular systems require an elaborate technological design. In addition to the basic radio component, a cellular system needs facilities for identifying the cell that contains an activated mobile unit, and for automatically switching duplex channel frequencies and transmitter stations as the mobile moves from cell to cell. When a caller wants to contact a mobile unit, the central processing computer for the location of the called mobile unit conducts a search. Once the unit's cell is identified, a signal is transmitted through wirelines to the appropriate base station; the trunking computer then assigns a channel in the proper cell, and the message is radioed to the mobile unit.

To deal with the problem of a mobile unit moving between cells, the idea of 'hand-off' was invented. The cellular system would be endowed with its own system-level switching and control capability; this is a higher layer in the mobile network, operating above the individual cells. Through continuous measurements of signal strength received from the individual cell-sites, the cellular system is able to sense when a mobile with a call in progress is passing from one cell to another, and to switch the call from the first cell to the second cell 'on the fly', without dropping or disrupting the call in progress. This requires fundamentally new techniques for determining which of several possible new cells the mobile has strayed into, as well as methods for tearing down and re-establishing the call in a very rapid manner. There are thus four principles that characterise cellular mobile telecommunications:
1. Lower-power transmitters and small coverage zones or cells
2. Frequency reuse
3. Cell-splitting to increase capacity
4. Hand-off and central control.

A2.2 The architecture of a cellular network

A mobile cellular telecommunications system has five main components: (1) radio base stations or air interface, (2) one or more switches to control them and route calls, (3) a subscriber database, (4) a telecommunications network that connects the base stations and switches with the public telecommunications network and finally (5) a mobile subscriber terminal. These are described in more detail below.

Base stations (air interface)

The base stations together form the radio system, or the air interface between the subscriber and the system. On top of carrying traffic, the radio system must continually monitor the position of the subscriber using signalling information so that the network can route the traffic to the base station within whose range the subscriber is located.

Each base station serves a cell and has its own master radio transmitter, receiver and antenna. Ideally cells are drawn as contiguous regular hexagons. In practice, however, cells are irregular and their shape varies, depending on the local topography. Moreover, cells must overlap to provide contiguous coverage.

As a mobile terminal crosses a cell boundary, a new channel must quickly be assigned so that uninterrupted communication can be maintained. The terminal equipment 'hands-off' the call from the base station in the user's original cell to the cell being entered. Components needed for the hand-off process, particularly the monitoring system that relays information back to the base control equipment and the switching circuits, are important determinants of service quality.

The way base stations are deployed depends on a series of parameters and in particular traffic requirements. If traffic is intense, such as in urban areas, base stations have to be deployed closer to each other; if traffic intensity is low, as in rural areas, base stations can be located further from each other. Other parameters concern the frequency of operation, the power level used and the efficiency with which frequencies can be reused.

Cellular systems cannot work with frequencies below 400 MHz – the signals would travel too far for reusing frequencies. Remember that the attenuation of the signal increases with frequency, affecting both the maximum and minimum feasible cell sizes. For instance, 450 MHz systems are suitable for rural areas, but not for urban areas with intense traffic because the minimum cell radius cannot go below 2 km. Likewise, 1800 MHz systems are good for urban areas but are not justified for rural areas. The maximum cell size of these systems is about 7 km, base stations would thus have to be spread densely which is not justified in rural areas because of the low traffic volume.

Power-level control is essential for the efficient working of a cellular system. This consists in limiting the power emitted by both base station and user terminal. In small cells power is then less, and in larger cells power can be increased to adjust to the greater distance between base station and user terminal. It is thanks to power-level control that the cellular network can work with cells of various size in order to cater for the different traffic volumes in urban areas.

To accommodate the maximum traffic, cells have to be as small as possible to achieve the maximum reuse of channels. The basic limiting factor of any cellular network is the level of interference that can be tolerated between transmitters on the same frequency as indicated by the SNR ratio. The lower this level, the closer the distance at which it is possible to use the same frequency. SNR ratios vary across cellular systems.

Several patterns for frequency repetition have been used. The smallest possible reuse pattern of non-adjacent hexagonal cells is made with four different cells. Other common patterns are the seven-cell patterns (where two cells are between cells of the same frequency) and the twelve-cell pattern (made with three cells in between). The level of interference that can be tolerated determines the choice of the pattern: the larger the repeat pattern, the further away are cells with the same frequency and the less interference. However, the greater the repeat pattern, the lower the capacity of the system.[2]

The general structure of a base station consists of a base station control function and a number of transceivers, commanding the entire area of coverage. The transceivers are controlled by the base station controller. The most important role of the base station controller is the management of frequencies used by all connected transceivers and the coordination of the hand-over.

The mobile switching centre

One part of the task of the mobile switching centre (MSC) is similar to that of a fixed telecommunications exchange: routing calls between subscribers. Indeed many of the switches used in cellular networks are derived from versions originally designed for fixed networks. However, the MSC has important additional tasks to perform. The most challenging is due to the fact that mobile telecommunications are moving, and so the MSC must find out where the user is in order to route calls appropriately as well as performing hand-over between cells whenever necessary. Because

---

[2] This can be illustrated by an example. Suppose eighty-four channels are available. A twelve-cell repeat pattern would enable the use of seven channels per base station, but a four-cell repeat pattern would allow the number of channels in each cell to be increased to twenty-one. Of course, it would be possible to increase the capacity of the twelve-cell system to an equivalent of the four-cell system by reducing the cell size. This, however, would require three–four times as many base stations. This illustrates the importance in reducing the interference between base stations to allow more efficient frequency reuse.

of this additional complexity an MSC typically handles a smaller number of subscribers than a fixed line exchange.

With cellular networks covering only a small area (e.g. a small country), there is only one switch. If, however, the network becomes larger, there may be several switches. Apart from optimising the workload, increasing the number of switches also helps reduce the use of the fixed line network and thus interconnection costs. Interconnection costs are typically set up in such a way that the further a call is carried, the higher is the interconnection fee. Mobile operators thus try to carry the call on its own network as far as possible. In this case, MSCs have also the task of routing the outgoing calls as economically as possible as well as enabling hand-over of calls between cells connected to different switches. For incoming calls, the task is even more complex, because the network must track the user before deciding the appropriate route and base station for the call. For this, the MSC must interrogate the subscriber database.

The subscriber database is the key component of a cellular system. Its task is two-fold: first, mobility management (i.e. maintaining a record of each subscriber and its position); second, call authentication. Mobility management uses the information generated by the control channels in the radio system. The database keeps a record of the base station from which a subscriber will receive the best signal. Whenever the mobile terminal is handed over from one cell to another, the record is updated. Incoming calls can then be routed appropriately.

The authentication process is necessary because a radio channel can be accessed by anyone transmitting on the correct frequency. Authentication establishes the validity of any mobile handset which tries to make a call on the system and of any limitation on the calls it is allowed to make. For instance, a mobile may be prevented from making international calls. This kind of information is stored in the subscriber database and interrogated when required, usually when the subscriber attempts to make a call. There are two registers called the home location register (HLR) for subscribers of the same network and the visitors' location register for subscribers coming from other networks (VLR), respectively.

Fixed transmission network

The fixed transmission network connects the different elements of the mobile network. This can be done either by fixed microwave links or by wirelines, either owned by the mobile operator or by another company. In the past it was quite common for the mobile operators to be subject to restrictions on owning a fixed transmission network and they therefore had to rely on leased lines provided by the monopoly fixed network operator. The design of the mobile network was much affected by the extent to which such obligations were imposed on the mobile operator. The system was not only designed to optimise capacity, but also carefully to balance leased line costs and interconnection payments. For instance, far-end hand-over reduces interconnection payments but increases leased line costs if additional capacity becomes necessary.

The mobile terminal

The mobile terminal or handset not only has to transmit and receive traffic, it also has to provide the authentication information required by the system before a call can be authorised and charged. Moreover, it has to send dialling information so that calls can be routed and has to communicate regularly with the system over a control channel so that its location can be monitored and its frequency changed whenever the system instructs it to hand-over from one base station to another. The mobile terminal must be able to operate on any of the radio channels allocated to the system, which means that it has to incorporate a frequency synthesiser. The mobile terminal has seen dramatic falls in price as a result of the progresses made in microelectronics. Technological improvements and price reductions have been one of the major factors of the rapid development of cellular mobile telecommunications.

A2.3 Functions specific to cellular networks

Whereas in a fixed telecommunications network the location of the user terminal is always known, with a mobile network this is not so. The mobile network thus requires a means of tracking a user who is being called. This function is called location management. To avoid interruption of a call when the user is moving from one cell to another, there must be 'hand-over' As cells size become smaller the frequency of hand-over increases and the smooth operation of this feature becomes more important.

'Roaming' refers to the possibility that a network is also able to host subscribers from a different network. A user therefore uses 'roaming' facilities when she is able to communicate via a network other than her own. This can be abroad (international 'roaming') or within the country of origin where there is no coverage by her own network (national 'roaming'). Besides issues of standard (i.e. networks must use compatible technical systems), it is necessary that the network operators in question enter agreements with each other.

A mobile call can be made to another mobile on the same network, to a mobile on a different network, or to a fixed line telephone. The way the mobile call is delivered to each of these destinations depends on both regulatory and commercial considerations. For example, a mobile to mobile (MTM) call on the same network could remain within the mobile operator's network if the operator is allowed to install the connecting infrastructure between the MSCs. Otherwise the call would go to the public fixed telecommunications network, to be routed to the mobile subscriber. Calls to a mobile telephone on different networks and calls to fixed lines can be routed directly or via the public fixed telecommunications network.

Bibliography

ACA, 2000. 'Telecommunications Performance Report 1999–2000', Australian Communications Agency

AGCM, 2001. 'Gara UMTS', Provvedimento n. 1445 of 27 June 2001, Rome: Autorità Garante della Concorrenza (Italian Competition Authority)

Ahn, H. and Lee, M., 1999. 'An econometric analysis of the demand for access to mobile telephone networks', *Information Economics and Policy*, 11: 297–305

Appleby, M. S., 1991. 'The UK cellular system', in R. C. V. Macario (ed.), *Personal and Mobile Radio Systems*, London: Peter Peregrinus

Armstrong, M., 1997. 'A simple model of competition in mobile telephony', London Business School, mimeo

1998. 'Network interconnection in telecommunications', *Economic Journal*, 108: 545–64

2002. 'The theory of access pricing and interconnection', in M. Cave, S. K. Majumdar and I. Vogelsang (eds.), *Handbook of Telecommunications Economics*, 1, Amsterdam: North-Holland

Armstrong, M., Cowan, S. and Vickers, J., 1994. *Regulatory Reform: Economic Analysis and British Experience*, Cambridge, MA: MIT Press

Armstrong, M. and Vickers, J., 1999. 'Competitive price discrimination', Oxford University, mimeo

Arthur, W. B., 1989. 'Competing technologies, increasing returns, and lock-in by historical events', *Economic Journal*, 99: 116–31

Artis, M. and Nixson, F., 2001. *The Economics of the European Union*, Oxford: Oxford University Press

Atkinson, J. M. and Barnekov, C. C., 2000. 'A competitively neutral approach to network interconnection', FCC, OPP Working Paper Series, 34

Ausubel, L. M. and Cramton, P., 1996. 'Demand reduction and inefficiency in multi-unit auctions', University of Maryland, mimeo

Ausubel, L. M., Cramton, P., McAfee, R. P. and McMillan, J., 1997. 'Synergies in wireless telephony: evidence from the broadband PCS auctions', *Journal of Economics & Management Strategy*, 6: 497–527

Ayres, I. and Cramton, P., 1996. 'Deficit reduction through diversity: a case study of how affirmative action at the FCC increased auction competition', *Stanford Law Review*, 48: 761–815

Bass, F. M., 1969. 'A new product model for consumer durables', *Management Science*, 15: 215–27

Bekkers, R. and Smits, J., 1997. *Mobile Telecommunications: Standards, Regulation and Applications*, Norwood, MA: Artech House

Berlage, M. and Schnöring, T., 1995. 'The introduction of competition in German mobile communication markets', in K. E. Schenk, J. Müller and T. Schnöring (eds.), *Mobile Telecommunications: Emerging European Markets*, Norwood, MA: Artech House

Bewley, R. and Fiebig, D. G., 1988. 'A flexible logistic growth model with applications in telecommunications', *International Journal of Forecasting*, 4: 177–92

Binmore, K. and Klemperer, P., 2002. 'The biggest auction ever: the sale of the British 3G licenses', *Economic Journal*, 112: 74–96

Björkdahl, J. and Bohlin, E., 2004. 'Competition policy and scenarios for European 3G markets', *Communications and Strategies*, 51: 21–34

Blackstone, E. and Ware, H., 1978. 'The cost of systems at 900 MHz', in R. Bowers, M. A. Lee and C. Hershey (eds.), *Communications for a Mobile Society: An Assessment of New Technology*, Beverly Hills, CA: Sage

Bonanno, G., 1985. 'Topics in oligoply', unpublished PhD thesis, London School of Economics

Börgers, T. and Dustmann, C., 2001. 'Strange bids: bidding behaviour in the United Kingdom's third generation spectrum auction', University College London, mimeo

  2002 'Rationalising the UMTS spectrum bids: the case of the UK auction', University College London, mimeo

Brock, G. W., 2002. 'Historical overview', in M. Cave, S. K. Majumdar and I. Vogelsang (eds.), *Handbook of Telecommunications Economics*, 1, Amsterdam: North-Holland

Brynjolfson, E. and Kemerer, C., 1996. 'Network externalities in microcomputer software: an econometric analysis of the spreadsheet market', *Management Science*, 42: 1627–47

Budd, C., Harris, C. and Vickers, J. S., 1993. 'A model of the evolution of duopoly: does the asymmetry between firms tend to increase or decrease?', *Review of Economic Studies*, 60: 543–74

Buehler, S. and Haucap, J., 2003. 'Mobile number portability', Discussion Paper, 17, University of Federal Armed Forces, Hamburg

Busse, M. R., 2000. 'Multimarket contact and price coordination in the cellular telephone industry', *Journal of Economics & Management Strategy*, 9: 287–320

Bylinsky, G., 1973. 'How Intel won its bet on memory chips', *Fortune*, November: 142–86

Cadima, N. and Pita Barros, P. L., 1998. 'The impact of mobile phone diffusion on the fixed line network', CEPR Working Paper, 2598, October

Calhoun, G., 1988. *Digital Cellular Radio*, Norwood, MA: Artech House

Cambini, C., Ravazzi, P. and Valletti, T., 2003. *Il mercato delle telecomunicazioni: dal monopolio alla liberalizzazione negli Stati Uniti e nella Unione Europea*, Bologna: Il Mulino

Cave, M. and Williamson, P., 1994. 'Entry, competition, and regulation in UK telecommunications', *Oxford Review of Economic Policy*, 12: 113–27

Chakravorti, B., Sharkey, W. W., Spiegel, J. and Wilkie, S., 1995. 'Auctioning the airwaves: the contest for broadband PCS spectrum', *Journal of Economics & Management Strategy*, 4: 345–74

Chow, G., 1967. 'Technological change and the demand for computers', *American Economic Review*, 57: 1117–30

Coase, R. H., 1959. 'The Federal Communications Commission', *Journal of Law and Economics*, 2: 1–40

Competition Commission, 2003. *Report on Charges Made by Mobile Operators for Terminating Calls*, London: HMSO

Cortada, J. W., 1987. *Historical Dictionary of Data Processing: Technology*, New York: Greenwood Press

Cramton, P., 1995. 'Money out of thin air: the nationwide narrowband PCS auction', *Journal of Economics & Management Strategy*, 4: 267–343

    1997. 'The FCC spectrum auction: an early assessment', *Journal of Economics & Management Strategy*, 6: 431–95

    2001. 'Lessons learned from the UK 3G spectrum auction', in National Audit Office, *The Auction of Radio Spectrum for the Third Generation of Mobile Telephones*, London: NAO

    2002. 'Spectrum auctions', in M. Cave, S. K. Majumdar and I. Vogelsang (eds.), *Handbook of Telecommunications Economics*, 1, Amsterdam: North-Holland

Cramton, P. and Schwartz, J. A., 2000. 'Collusive bidding: lessons from FCC spectrum auctions', *Journal of Regulatory Economics*, 17: 229–52

Crandall, R. W., 1998. 'New Zealand spectrum policy: a model for the United States?', *Journal of Law and Economics*, 41: 821–40

Cremer, H., Ivaldi, M. and Turpin, E., 1996. 'Competition in access technologies', Working Paper, DT 60, IDEI, Université de Toulouse

Curwen, P., 2002. *The Future of Mobile Communications: Awaiting the Third Generation*, Basingstoke: Palgrave

Dana, J. D. and Spier, K. E., 1994. 'Designing a private industry: government auctions with endogenous market structure', *Journal of Public Economics*, 53: 127–47

Davies, S., 1979. *The Diffusion of Process Innovations*, Cambridge: Cambridge University Press

Dayem, R. A., 1997. *PCS and Digital Cellular Technologies: Assessing your Options*, Upper Saddle River, NJ: Prentice-Hall

DeGraba, P., 2000. 'Bill and keep at the central office as the efficient interconnection regime', FCC, OPP Working Paper Series, 33

    2002. 'Efficient inter-carrier compensation for competing networks when customers share the value of a call', FTC, mimeo

Dekimpe, M. G., Parker, P. M. and Sarvary, M., 1998. 'Staged estimation of international diffusion models: an application to global cellular telephone adoption', *Technological Forecasting and Social Change*, 57: 105–32

Didier, M. and Lorenzi, J.-H., 2002. *Enjeux economiques de l'UMTS*, Paris: La Documentation Française

Dixon, R., 1980. 'Hybrid corn revisited', *Econometrica*, 48: 1451–61

Doyle, C. and Smith, J. C., 1998. 'Market structure in mobile telecoms: qualified indirect access and the receiver pays principle', *Information Economics and Policy*, 10: 471–88

Dresdner Kleinwort Wasserstein, 2002. 'European mobile operators: mixed signals', Research Report, June

Duso, T., 2000. 'Who decides to regulate? Lobbying activity in the US cellular industry', WZB Discussion Paper, FS IV 00–05

Economic Commission for Europe, 1987. *The Telecommunications Industry: Growth and Structural Change*, New York: United Nations

Economides, N., 1996. 'The economics of networks', *International Journal of Industrial Organization*, 14: 673–99

European Commission, 1994. 'Towards the personal communications environment: Green Paper on a common approach in the field of mobile and personal communications', COM(94) 145 final

    1997a. 'On the further development of mobile and wireless communications', COM(97) 217 final

    1997b. '3rd implementation report of the telecommunications regulatory package', COM(1997) 80 final

    2000. 'On the initial findings of the sector inquiry into mobile roaming charges', Working Document, DG Competition, 13 December

    2001. 'The introduction of 3G mobile communications in the EU: state of play and the way forward', COM(2001) 141 final

    2002a. 'Comparative assessment of the licensing regimes for 3G mobile communications in the European Union and their impact on the mobile communications sector', Final Report, June

    2002b. 'Towards the full roll-out of third generation mobile communications', COM(2002) 301 final

    2002c. '8th implementation report of the telecommunications regulatory package', COM(2002) 695 final

Ewerhard, C. and Moldovanu, B., 2002. 'A stylised model of the German UMTS auction', University of Mannheim, mimeo

Farrell, J. and Saloner, G., 1985. 'Standardization, compatibility, and innovation', *The RAND Journal of Economics*, 16: 70–83

    1986. 'Installed base and compatibility: innovation, product preannouncements, and predation', *American Economic Review*, 76: 940–55

Farrell, J. and Shapiro, C., 1992. 'Standard setting in high-definition television', *Brooking Papers on Economic Activity: Microeconomics*: 1–77

Federal Communications Commission (FCC), 1995. 'First annual report and analysis of competitive market conditions with respect to commercial mobile services', FCC 95–319, Washington, DC: Federal Communications Commission

    1997. 'The FCC report to Congress on spectrum auctions', FCC Wireless Telecommunications Bureau, WT Docket, 97–150

2000. 'Fifth annual report and analysis of competitive market conditions with respect to commercial mobile services', FCC 00–289, Washington, DC: Federal Communications Commission

2003. 'Eighth annual report and analysis of competitive market conditions with respect to commercial mobile services', FCC 03–150, Washington, DC: Federal Communications Commission

Foreman, R. D. and Beauvais, E., 1999. 'Scale economies in cellular telephony: size matters', *Journal of Regulatory Economics*, 16: 297–306

Frey, J. and Lee, A. M., 1978. 'Technologies for land mobile communications: 900 MHz systems', in R. Bowers, A. M. Lee and C. Hershey (eds.), *Communications for a Mobile Society*, Beverly Hills, CA: Sage

Funk, J. L., 1998. 'Competition between regional standards and the success and failure of firms in the worldwide mobile communications market', *Telecommunications Policy*, 22: 419–41

2002. *Global Competition between and within Standards: The Case of Mobile Phones*, Basingstoke: Palgrave

Gabszewicz, J. J. and Thisse, J. F., 1979. 'Price competition, quality and income disparities', *Journal of Economic Theory*, 20: 340–59

1980. Entry (and exit) in a differentiated industry', *Journal of Economic Theory*, 22: 327–38

Gandal, N., 1994. 'Hedonic price indexes for spreadsheets and an empirical test for network externalities', *RAND Journal of Economics*, 25: 160–70

2002. 'Compatibility, standardisation and network effects: some policy implications', *Oxford Review of Economic Policy*, 18: 80–91

Gans, J. S. and King, S. P., 1999. 'Termination charges for mobile phone networks: competitive analysis and regulatory options', University of Melbourne, mimeo

2000. 'Mobile network competition, customer ignorance, and fixed-to-mobile call prices', *Information Economics and Policy*, 12: 301–28

Gans, J. S., King, S. P. and Woodbridge, G., 2001. 'Numbers to the people', *Information Economics and Policy*, 13, 167–80

Garg, V. K. and Wilkes, J. E., 1996. *Wireless and Personal Communications Systems*, Upper Saddle River, NJ: Prentice-Hall

Garrard, G. A., 1998. *Cellular Communications: Worldwide Market Developments*, Norwood, MA: Artech House

Gerardin, D. and Kerf, M., 2003. *Controlling Market Power in Telecommunications: Antitrust vs. Sector-specific Regulation*, Oxford: Oxford University Press

Geroski, P., 2000. 'Models of technology diffusion', *Research Policy*, 29: 603–25

Geroski, P., Thomson, D. and Tooker, S., 1989. 'Vertical separation and price discrimination: cellular phones in the UK', *Fiscal Studies*, 10: 83–103

Griliches, Z., 1957. 'Hybrid corn: an exploration in the economics of technical change', *Econometrica*, 25: 501–22

Grimm, V., Riedel, F. and Wolfstetter, E., 2001. 'The third generation (UMTS) spectrum auction in Germany', CESifo Working Paper, 584

2002. 'Low price equilibrium in multi-unit auctions: the GSM spectrum auction in Germany', Humboldt Universität, Berlin, mimeo

2003. 'Implementing efficient market structure: optimal licensing in natural oligopoly when tax revenue matters', *Review of Economic Design*, 7: 443–63

Grindley, P., 1995. *Standards, Strategy and Policy: Cases and Stories*, Oxford: Oxford University Press

Gruber, H., 1994. *Learning and Strategic Product Innovation: Theory and Evidence for the Semiconductor Industry*, Amsterdam: North-Holland

1998. 'The diffusion of innovations in protected industries: the textile industry', *Applied Economics*, 30: 77–83

1999. 'An investment view of mobile telecommunications in the European Union', *Telecommunications Policy*, 23: 521–38

2001. 'Competition and innovation: the diffusion of mobile telecommunications in Central and Eastern Europe', *Information Economics and Policy*, 13: 19–34

2002. 'Endogenous sunk costs in the market for mobile telecommunications: the role of licence fees', *The Economic and Social Review*, 33: 55–64

2004. 'Radio spectrum fees as determinants of market structure: the consequences of European 3G licensing', in L. Soete and B. ter Weel (eds.), *The Economics of the Digital Economy*, Cheltenham: Edward Elgar, forthcoming

Gruber, H. and Hoenicke, M., 1998. 'The European mobile telecommunications market', European Investment Bank, mimeo

1999. 'The road toward third generation mobile telecommunications', *Info*, 3: 213–24

2000. 'Third generation mobile. what are the challenges ahead?', *Communications and Strategies*, 39: 159–73

Gruber, H. and Valletti, T., 2003. 'Mobile telecommunications and regulatory frameworks', in G. Madden and S. Savage (eds.), *The International Handbook of Telecommunication Economics, II: Emerging Telecommunications Networks*, Cheltenham: Edward Elgar

Gruber, H. and Verboven, F., 2001a. 'The diffusion of mobile telecommunications services in the European Union', *European Economic Review*, 45: 577–88

2001b. 'The evolution of markets under entry and standards regulation – the case of global mobile telecommunications', *International Journal of Industrial Organisation*, 19: 1189–1212

Guerci, C. M., Cervigni, G., Marcolongo, G. and Pennarola, F., 1998. *Monopolio e concorrenza nelle telecomunicazioni: Il caso Omnitel*, Milan: Il Sole 24 Ore

Hamilton, J., 2003. 'Are mainlines and mobile phones substitutes or complements? Evidence from Africa', *Telecommunications Policy*, 27: 190–233

Hausman, J. A., 1997. 'Valuing the effect of regulation on new services in telecommunications', *Brooking Papers on Economic Activity: Microeconomics*: 1–38

2000. 'Efficiency effects on the US economy from wireless taxation', *National Tax Journal*, 53: 733–42

2002. 'Mobile Phone', in M. Cave, S. K. Majumdar and I. Vogelsang (eds.), *Handbook of Telecommunications Economics, 1*, Amsterdam: North-Holland

Hazlett, T. W., 1998. 'Assigning property rights to radio spectrum users: why did the FCC licence auctions take 67 years?', *Journal of Law and Economics*, 41: 529–75

Hazlett, T. W. and Boliek, B. E. L., 1999. 'Use of designated entry preferences in assigning wireless licenses', *Federal Communications Law Journal*, 51: 639–62

Hazlett, T. W. and Michaels, R. J., 1993. 'The cost of rent-seeking: evidence from cellular telephone license lotteries', *Southern Journal of Economics*, 59: 425–9

Hotelling, H., 1929. 'Stability in competition', *Economic Journal*, 39: 41–57

Hultén, S., Andersson, P. and Valiente, P., 2001. '3G mobile policy: the case of Sweden', Stockholm School of Economics, mimeo

International Telecommunications Union (ITU) 1999. *World Telecommunication Development Report 1999. Mobile Cellular*, Geneva: International Telecommunications Union

    2002a. *World Telecommunication Development Report 2002. Reinventing Telecoms*, Geneva: International Telecommunications Union

    2002b. *Trends in Telecommunications Reform 2002*, Geneva: International Telecommunications Union

    2003a. *Mobile Overtakes Fixed: Implications for Policy and Regulation*, Geneva: International Telecommunications Union

    2003b. 'Broadband Korea: Internet case study', March, mimeo

International Trade Commission (ITC), 1993. 'Global competitiveness of US advanced-technology industries: cellular communications', Publication 2646, Washington, DC: International Trade Commission

Jeon, D., Laffont, J. and Tirole, J., 2001. 'On the receiver pays principle', Pompeu Fabra University, mimeo

Jha, R. and Majumdar, S. K., 1999. 'A matter of connections: OECD telecommunications sector productivity and the role of cellular technology diffusion', *Information Economics and Policy*, 11: 243–69

Jung, V. and Warnecke, H.-J., 1998. *Handbuch für die Telekommunikation*, Berlin: Springer

Kahn, A. E. 2004. *Lessons from Deregulation: Telecommunications and Airlines After the Crunch*, Washington, DC: Brookings Institution Press

Kargman, H., 1978. 'Land mobile communications: the historical roots', in R. Bowers, A. M. Lee and C. Hershey (eds.), *Communications for a Mobile Society*, Beverly Hills, CA: Sage

Katz M. L. and Shapiro, C., 1985. 'Network externalities, competition, and compatibility', *American Economic Review*, 77: 402–20

    1986. 'Technology adoption in the presence of network externalities', *Journal of Political Economy*, 94: 822–41

    1994. 'Systems competition and network effects', *Journal of Economic Perspectives*, 8: 93–115

Kim, H. S. and Kwon, N., 2002. 'The advantage of networks size in acquiring new subscribers: a conditional logit analysis of the Korean mobile telephony market', *Information Economics and Policy*, 15: 17–33

Kim, J. Y. and Lim, Y., 2001. 'An economic analysis of the receiver pays principle', *Information Economics and Policy*, 13: 231–60

King, J. L. and West, J., 2002. 'Ma Bell's orphan: US cellular telephony, 1947–1996', *Telecommunications Policy*, 26: 189–203

Klemperer, P., 1989. 'Price wars caused by switching costs', *Review of Economic Studies*, 56: 405–20

1995. 'Competition when consumers have switching costs: an overview with applications to industrial organization, macroeconomics, and international trade', *Review of Economic Studies*, 62: 515–39

1999. 'Auction theory: a guide to the literature', *Journal of Economic Surveys*, 13. 227–86

2002a. 'How (not) to run auctions: the European 3G telecom auctions', *European Economic Review*, 46: 829–45

2002b. 'Using and abusing economic theory', Oxford University, mimeo

2002c. 'What really matters in auction design', *Journal of Economic Perspectives*, 16: 169–90

Laffont, J.-J. and Tirole, J., 2000. *Competition in Telecommunications*, Cambridge, MA: MIT Press

Lancaster, K., 1966. 'A new approach to consumer theory', *Journal of Political Economy*, 74: 796–821

Lane, W., 1980. 'Product differentiation in a market with endogenous sequential entry', *Bell Journal of Economics*, 11: 237–70

Levin, H. J., 1971. *The Invisible Resource: Use and Regulation of the Radio Spectrum*, Baltimore, MD: Johns Hopkins Press

Levin, S. G., Levin, S. L. and Meisel, J. B., 1987. 'A dynamic analysis of the adoption of a new technology: the case of optical scanners', *Review of Economics and Statistics*, 69: 12–17

Levine, M. E., 1998. 'Regulatory capture', in P. Newman (ed.), *New Palgrave Dictionary of Economics and the Law*, 3, Basingstoke: Palgrave

Liebowitz, S. J. and Margolis, S. E., 1999. *Winners, Losers & Microsoft: Competition and Antitrust in High Technology*, Oakland, CA: The Independent Institute

Lüngen, B., 1985. 'The role of mobile telecommunications in Central-Eastern Europe', K. E. Schenk, J. Müller and T. Schnöring (eds.), *Mobile Telecommunications: Emerging European Markets*, Norwood, MA: Artech House

Mahajan, V., Muller, E. and Bass, F. M., 1993. 'New product diffusion models', in J. Eliashberg and G. L. Lilien (eds.), *Handbook of Operational Research and Management Science*, 5, Amsterdam: North-Holland

Malerba, F., 1985. *Semiconductor Business: The Economics of Rapid Growth and Decline*, Madison, WI: University of Wisconsin Press.

Manguian, J. P., 1993. *Les radiocommunications*, Paris: Presses Universitaires Françaises

Matutues, C. and Regibeau, P., 1988. 'Mix and match: product compatibility without network externalities', *The RAND Journal of Economics*, 19: 221–34

McAfee, R. P. and McMillan, J., 1996. 'Analyzing the airwaves auction', *Journal of Economic Perspectives*, 10: 159–76

McAfee, R. P. and Vincent, D., 1983. 'The declining price anomaly', *Journal of Economic Theory*, 60: 191–212

McKenzie, D. J. and Small, J. P., 1997. 'Econometric cost structure for cellular telephony in the United States', *Journal of Regulatory Economics*, 12: 147–57

McMillan, J., 1994. 'Selling spectrum rights', *Journal of Economic Perspectives*, 8: 145–62

Mehrotra, A., 1994. *Cellular Radio: Analog and Digital Systems*, Norwood, MA: Artech House

Melody, W. H., 2001. 'Spectrum auctions and efficient resource allocation: learning from the 3G experience in Europe', *Info*, 3: 5–10

Milgrom, P. and Weber, R. J., 1982. 'A theory of auctions and competitive bidding', *Econometrica*, 50: 1089–1122

Ministry of Commerce of New Zealand, 1995. 'Radiocommunications Act review: preliminary conclusions', December

Ministry of Economic Development of New Zealand, 2000. 'New Zealand spectrum management. a decade in review: 1989–1999', PIB, 35, June

Ministry of Transport and Communications Finland, 1998. 'Telecommunications statistics 1998', Helsinki: Finnish Ministry of Transport and Communications

Miravete, E. J. and Röller, L. H., 2003. 'Competitive non-linear pricing in duopoly equilibrium: the early US cellular telephone industry', CEPR Discussion Paper, 4069

Mitchell, I. and Vogelsang, B. M., 1997. *Telecommunications Competition: The Last Ten Miles*, Cambridge, MA: MIT Press

MMC, 1998. 'Cellnet and Vodafone', Joint publication by the UK Monopolies and Merger Commission and Oftel, London

Mölleryd, B. G., 1997. 'The building of a world industry: the impact of entrepreneurship on Swedish mobile telephony', Teldok, 28e, Stockholm

Morand, P.-H., Mougeot, M. and Naegelen, F., 2001. 'UMTS: fallait-il choisir un concours de beauté?', *Revue d'économie politique*, 5: 669–82

Moreton, P. S. and Spiller, P. T., 1998. 'What's in the air: interlicence synergies in the Federal Communications Commission's broadband personal communications service spectrum auctions', *Journal of Law and Economics*, 41: 677–716

Morris, P. R., 1990. *A History of the World Semiconductor Industry*, London: Peter Peregrinus

Mowery, D. C. and Rosenberg, N., 1998. *Paths of Innovation: Technological Change in 20th Century America*, Cambridge: Cambridge University Press

Müller, J. and Callmer, C., 1995. 'The role of international players in mobile telecommunications in Central and Eastern Europe', in K. E. Schenk, J. Müller and T. Schnöring (eds.), *Mobile Telecommunications: Emerging European Markets*, Norwood, MA: Artech House

Müller J. and Tooker, S., 1994. 'Mobile communications in Europe', in C. Steinfield, J. M. Bauer and L. Caby, *Telecommunications in Transition: Policies, Services and Technologies in the European Community*, Thousand Oaks, CA: Sage

NAO, 2001. 'The auction of radio spectrum for the third generation of mobile telephones', London: NAO, November

Nattermann, P. M., 1999. 'Estimating firm conduct: the German cellular market', Doctoral thesis, Georgetown University

Nattermann, P. M. and Murphy, D. D., 1998. 'The Finnish telecommunications market: advantage of local access incumbency', *Telecommunications Policy*, 22: 757–73

NERA, 1988. 'Management of the radio frequency spectrum in New Zealand', National Economic Research Associates, mimeo, November

Noam, E., 1992. *Telecommunications in Europe*, New York: Oxford University Press

Noble, D., 1962. 'The history of land mobile communications', *Proceedings of the IRE, Vehicular Communications*: 157–79

OECD, 1993. *Communications Outlook*, Paris: Organisation for Economic Cooperation and Development

   1996a. *Mobile Cellular Communication: Pricing Strategy and Competition*, Paris: Organisation for Economic Cooperation and Development

   1996b. 'OECD reflection on the benefits of mobile cellular telecommunications infrastructure competition', GD(96)42, Paris: Organisation for Economic Cooperation and Development

   1997. *The OECD Report on Regulatory Reform, 1: Sectoral Studies*, Paris: Organisation for Economic Cooperation and Development

   1999. 'Regulatory reform in Japan: regulatory reform in the telecommunications industry', Paris: Organisation for Economic Cooperation and Development, mimeo

   2000. 'Cellular mobile pricing structures and trends', DSTI/ICCP/TISP(99)11/final, Paris: Organisation for Economic Cooperation and Development

   2001. *Communications Outlook*, Paris: Organisation for Economic Cooperation and Development

   2003. *Communications Outlook*, Paris: Organisation for Economic Cooperation and Development

Offerman, T. and Potters, J., 2000. 'Does auctioning of entry licences affect consumer' prices? An experimental study,' CENter working paper, 55, Tilburg University

Oftel, 1999a. 'Statement on mobile virtual network operators', London: Office of Telecommunications

   1999b. 'Statement on national roaming', London: Office of Telecommunications

   2001a. 'Effective competition review: mobile', London: Office of Telecommunications, February

   2001b. 'Effective competition review: mobile', London: Office of Telecommunications, September

   2002. 'Vodafone, O$_2$, Orange and T-mobile', London: Office of Telecommunications and Monopolies and Mergers Commission, December

Olmsted Teisberg, E., 1992. 'Technology choice in digital cellular phone switches', Working Paper, 92–062, Harvard Business School

Oster, S., 1982. 'The diffusion of innovation among steel firms: the basic oxygen furnace', *Bell Journal of Economics*, 13: 45–56

Palmberg C., 1998. 'Industrial transformation through public technology procurement? The case of the Finnish telecommunications industry', Doctoral thesis, Åbo University

Parker, P. M. and Röller, L.-H., 1997. 'Collusive conduct in duopolies: multi-market contact and cross-ownership in the mobile telephone industry', *The RAND Journal of Economics*, 28: 304–22

Pisjak, P., 1995. 'Mobile telecommunications in Austria', in K. E. Schenk, J. Müller and T. Schnöring (eds), *Mobile Telecommunications: Emerging European Markets*, Norwood, MA: Artech House

Plott, C. and Salmon, T., 2001. 'The simultaneous, ascending auction: dynamics of price adjustment in experiments and in the field', California Institute of Technology and Florida State University, mimeo

PNE, 2000. *Mobile Yearbook*, London, The Economist Group

Prestowitz, C. V., Jr, 1988. *Trading Places: How We Are Giving Our Future to Japan and How to Reclaim It*, New York: Basic Books

Pulkkinen, M., 1997. 'The breakthrough of Nokia mobile phones', Doctoral dissertation, Helsinki School of Economics and Business, A-122

Rappaport, T. S., 1996. *Wireless Communications: Principles and Practice*, Upper Saddle River, NJ: Prentice-Hall

Ray, G. F. (ed.), 1984. *The Diffusion of Mature Technologies*, Cambridge: Cambridge University Press

Redl, S. M., Weber, M. K. and Oliphant, M. W., 1995. *An Introduction to GSM*, Norwood, MA: Artech House

Regli, B. J. W., 1997. *Wireless: Strategically Liberalizing the Telecommunications Market*, Mahwah, NJ: Lawrence Erlbaum Associates

Reiffen, D., Schumann, L. and Ward, M. R., 2000. 'Discriminatory dealing with downstream competitors: evidence from the cellular industry', *Journal of Industrial Economics*, 48: 253–86

Reiffen, D. and Ward, M. R., 1997. 'Discriminatory dealing with downstream competitors: evidence from the cellular industry', FCC, mimeo

Reinganum, J., 1989. 'The timing of innovation: research, development and diffusion', in R. Schmalensee and R. Willig (eds.), *Handbook of Industrial Organization, I*, Amsterdam: North-Holland

Rochet, J.-C. and Stole, L., 2000. 'The economics of multidimensional screening', University of Chicago, mimeo

Rogers, E. M., 1995. *Diffusion of Innovations*, 4th edn., New York: Free Press

Rohlfs, J. H., 2001. *Bandwagon Effects in High-technology Industries*, Cambridge, MA: MIT Press

Rohlfs J. H., Jackson, C. L. and Kelley, T. E., 1991. 'Estimate of the loss to the United States caused by the FCC delay in licensing cellular telecommunications', National Economic Research Associates, November, mimeo

Rose, N. L. and Joskow, P. L., 1990. 'The diffusion of new technologies: evidence from the electric utility industry', *The RAND Journal of Economics*, 21: 354–73

Ruiz, L. K., 1995. 'Pricing strategies and regulatory effects in the US cellular telecommunications duopolies', in G. W. Brock (ed.), *Towards a Competitive Telecommunications Industry*, Mahwah, NJ: Lawrence Erlbaum Associates

Salomon Brothers, 1997. 'Wireless Europe', Broker report, March

Saloner, G. and Shepard, A., 1995. 'Adoption of technologies with network effects: an empirical analysis of the adoption of automated teller machines', *The RAND Journal of Economics*, 26: 479–501

Salop, S., 1979. 'Monopolistic competition with outside goods', *Bell Journal of Economics*, 10: 141–56

Salop, S. and Stiglitz, J., 1977. 'Bargains and ripoffs: a model of monopolistically competitive price dispersion', *Review of Economic Studies*, 44: 493–510

Schenk, K. E., Müller, J. and Schnöring, T. (eds.), 1995. *Mobile Telecommunications: Emerging European Markets*, Norwood MA: Artech House

Shaked, A. and Sutton, J., 1982. 'Relaxing price competition through product differentiation', *Review of Economic Studies*, 49: 3–13
    1983. 'Natural oligopolies', *Econometrica*, 51: 1469–83

Shapiro, C. and Varian, H. R., 1999. *Information Rules: A Strategic Guide to the Network Economy*, Boston, MA: Harvard Business School Press

Shew, W. B., 1994. 'Regulation, competition, and prices in the US cellular telephone industry', American Enterprise Institute, mimeo

Sidak, J. G. and Spulber, D. F., 1996. 'Deregulatory takings and breach of the regulatory contract', *New York University Law Review*, 71: 851–999

Song, J. D. and Kim, J. C., 2001. 'Is five too many? Simulation analysis of profitability and cost structure in the Korean mobile telephony industry', *Telecommunications Policy*, 25: 101–23

Steinbock, D., 2001. *The Nokia Revolution: The Story of an Extraordinary Company that Transformed an Industry*, New York: Amacom
    2003. *Wireless Horizon: Strategy and Competition in the Worldwide Marketplace*, New York: Amacom

Stoneman, P., 1983. *The Economic Analysis of Technological Change*, Oxford: Oxford University Press

Sutton, J., 1991. *Sunk Cost and Market Structure*, Cambridge, MA: MIT Press
    1998 *Technology and Market Structure: Theory and History*, Cambridge, MA: MIT Press

Taylor, L. D., 1994. *Telecommunications Demand in Theory and Practice*, Boston, MA: Kluwer

Temin, P., 1987. *The Fall of the Bell System*, New York: Cambridge University Press

Tengg, H., 1997. 'Liberalisierung der Mobilkommunikation in Österreich', in A. Kaspar and P. Rübig (eds.), *Telekommunikation: Herausforderung für Österreich*, Vienna: Signum

Tirole, J., 1988. *The Theory of Industrial Organization*, Cambridge, MA: MIT Press

Tyson, L. D., 1992. *Who's Bashing Whom? Trade Conflict in High Technology Industries*, Washington, DC: Institute for International Economics

US DOC, 2001. 'The potential for accommodating third generation mobile systems in the 1710–1850 MHz band: Federal operations, relocation costs and operational impacts', Final Report, Washington, DC: US Department of Commerce, March

Valletti, T. M., 1999. 'A model of competition in mobile communications', *Information Economics and Policy*, 11: 61–72

2000. 'Switching costs in vertically related markets', *Review of Industrial Organization*, 17: 305–409

2001. 'Spectrum trading', *Telecommunications Policy*, 25: 655–70

2003. 'Is mobile telephony a natural oligopoly?', *Review of Industrial Organization*, 22: 47–65

Valletti, T. M. and Cave, M., 1998. 'Competition in UK mobile communications', *Telecommunications Policy*, 22: 109–31

Van Damme, E., 2001. 'The Dutch auction in retrospect', mimeo, May

Van De Wielle, B. and Verboven, F., 2000. 'The timing of entry with consumer switching costs', University of Antwerp, mimeo

Vickrey, W., 1961. 'Counterspeculation, auctions, and competitive sealed tenders', *Journal of Finance*, 16: 32–55

Warburg Dillon Reed, 1998. 'The European cellular growth analyser', July

Webb, W., 1998. *Understanding Cellular Radio*, Norwood, MA: Artech House

Whalley, J. and Curwen, P., 2003. 'Licence acquisition strategy in the European mobile communications industry', *Info*, 5: 45–59

Withers, D., 1999. *Radio Spectrum Management: Management of the Spectrum and Regulation of Radio Services*, 2nd edn, London: Institution of Electrical Engineers

Wolfstetter, E., 2001. 'The Swiss UMTS spectrum auction flop: bad luck or bad design?', mimeo, May

World Bank, 1994. *World Development Report 1994: Infrastructure for Development*, Oxford: Oxford University Press

1998. 'Telecommunications strategy in the Europe and Central Asia region', mimeo, December

Wright, J., 1999. 'International telecommunications, settlement rates, and the FCC', *Journal of Regulatory Economics*, 15: 267–91

2000, 'Competition and termination in cellular networks', University of Auckland, mimeo

2002. 'Access pricing under competition: an application to cellular networks', *Journal of Industrial Economics*, 50: 289–315

Index

DATE DUE